Semiconductor Fundamentals

FIRST EDITION

BY PRASUN BARUA

ABOUT

Welcome to Semiconductor Fundamentals! This is a nonfiction science book which contains various topics on fundamentals of semiconductor. Semiconductors are a vital component of electronic equipment, allowing for advancements in communications, computers, healthcare, military systems, transportation, clean energy, and a variety of other fields. Semiconductors, also known as integrated circuits (ICs) or microchips, are produced from pure components such as silicon or germanium, as well as compounds like gallium arsenide. Small amounts of impurities are added to these pure elements in a process called doping, causing dramatic variations in the conductivity of the material. Semiconductors play a significant function in developing electronic gadgets, therefore they're a big part of our life. Consider what life would be like if we didn't have access to technology devices like smartphones, radios, televisions, laptops, video games and advanced medical diagnostic equipment. This book covers various topics on Basics of Semiconductor, PN Junction Theory, PN Junction Diode, The Signal Diode, Power Diodes and Rectifiers, Full Wave Rectifier, The Zener Diode, The Light Emitting Diode, Bypass Diodes in Solar Panels, Diode Clipping Circuits and The Schottky Diode. This is the first edition of the book. Thanks for reading the book.

TABLE OF CONTENTS

CHAPTER NO.	TITLE	PAGE NO.
CHAPTER-1	BASICS OF SEMICONDUCTOR	5
CHAPTER-2	THEORY OF PN JUNCTION	15
CHAPTER-3	PN JUNCTION DIODE	19
CHAPTER-4	THE SIGNAL DIODE	29
CHAPTER-5	POWER DIODES AND RECTIFIERS	39
CHAPTER-6	FULL WAVE RECTIFIER	46
CHAPTER-7	THE ZENER DIODE	58
CHAPTER-8	THE LIGHT EMITTING DIODE	68
CHAPTER-9	BYPASS DIODES IN SOLAR PANELS	85
CHAPTER-10	DIODE CLIPPING CIRCUITS	92
CHAPTER-11	THE SCHOTTKY DIODE	99

CHAPTER-1: BASICS OF SEMICONDUCTOR

What is Diode?

Diode is the most basic active semiconductor component in any electronic circuit. A diode contains exponential Current-Voltage (I-V) relationship. Current flows through diode in one direction only. That's why diodes are called basic unidirectional semiconductor devices. In Forward Biased Condition, it behaves same like a one-way electrical valve. A single piece of Semiconductor material is used to build a diode.

Diode has a positive "P-region" at one end and a negative "N-region" at the other, and which has a resistivity value somewhere between that of a conductor and an insulator.

What makes a "Semiconductor" material?

In order to understand semiconductor material, we have to understand about resistivity, conductor and insulator.

Resistivity

According to Ohm's Law, resistance is the proportion of the voltage difference to the flow of current across any electrical or electronic circuit. The issue with involving opposition as an estimation is that it relies especially upon the actual size of the material being estimated as well as the material out of which it is made. For instance, if we somehow managed to expand the length of the material (making it longer) its opposition would likewise increment relatively.

Moreover, assuming we expanded its width or size (making it thicker) its obstruction worth would diminish. So we need to have the option to characterize the material so as to demonstrate its capacity to one or the other lead or go against the progression of electrical flow through it regardless of anything else its size or shape is.

The quantity that is used to indicate this specific resistance is called Resistivity and is given the Greek symbol of ρ, (Rho). Resistivity is measured in Ohm-meters, (Ω.m). Resistivity is the inverse to conductivity.

If the resistivity of various materials is compared, they can be classified into three main groups, Conductors, Insulators and Semi-conductors as shown below chart of resistivity:

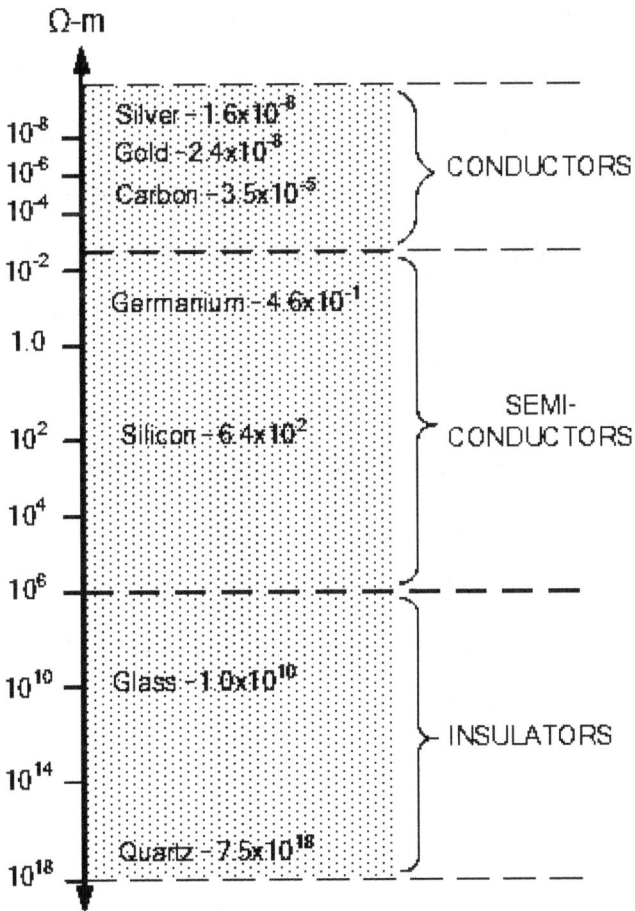

Here, a very little margin between the resistivity of the conductors like gold and silver in comparison to a much larger margin for the resistivity of the insulators between quartz and glass.

Because of the variation of ambient temperature, there is a difference in resistivity. Metals are much better conductors of heat than insulators.

Conductors

The resistivity of conductors is very little which is typically in the micro-ohms per meter. There are lots of free electrons in the basic atomic structure of a conductor which assists to flow electrical current conveniently. When electrical voltage is applied, these electrons will only flow through a conductor.

If a positive voltage potential is applied to the material, these "free electrons" leave their parent atom and travel together through the material forming an electron drift, more commonly known as a current. How "freely" these electrons can move through a conductor depends on how easily they can break free from their basic atoms when a voltage is applied. Then the amount of electrons that flow depends on the amount of resistivity the conductor has.

Metals like Copper, Aluminum, Silver or non-metals like Carbon are usually good conductor as they contain very few electrons in their outer "Valence Shell" or ring. It causes them being easily knocked out of the atom's orbit. This helps them to flow freely through the material until they intersect with other atoms, producing a "Domino Effect" through the material thereby creating an electrical current. Copper and Aluminum is the main conductor used in electrical cables as shown.

As metals contain very little values of resistance, they are good conductors of electricity. Typically, it is micro-ohms per meter, ($\mu\Omega.m$). Copper and aluminum are examples of very good conductors of electricity. They still have some resistance to the flow of electrons and subsequently do not conduct seamlessly. The resistivity of conductors increases with ambient temperature. Therefore, during carrying electrical current, resistors become hot.

Insulators

Insulators are built with non-metals materials which don't have any free electrons in their basic atomic structure. In this case, the electrons in the outer valence shell are intensely attracted by the positively charged inner nucleus. As there are no free electrons, no current will flow even though a potential voltage is applied to the material. It provides these materials their insulating properties.

Insulators are not affected by normal temperature changes because of having their very high resistivity. Typically, it is millions of ohms per meter. Although at very high temperatures wood becomes charcoal and changes from an insulator to a conductor. Some examples of good insulators are rubber, marble, fused quartz, PVC plastics etc.

Without insulators, electrical circuits would short together and not work. Therefore, insulators play a vital role in electrical and electronic circuits. For insulating and supporting overhead transmission cables, insulators made of glass or porcelain are used. Some other examples: epoxy-glass resin materials are used to make printed circuit boards, PCB's etc. while PVC is used to insulate electrical cables.

Semiconductor materials

Semiconductors materials like silicon (Si), germanium (Ge) and gallium arsenide (GaAs), contains electrical properties in between a "conductor" and an "insulator". They are neither good conductors nor good insulators. Therefore, they are called "semi"-conductors. Their atoms are closely assembled together in a crystalline form called a "crystal lattice". Therefore, they have very little "free electrons". Under special conditions, these electrons can flow.

By adding or replacing, certain donor or acceptor atoms to this crystalline structure, the ability of semiconductors to conduct electricity can be significantly developed. In this way, thereby, more free electrons can be produced than holes or vice versa. That is by adding a little percentage of another element to the base material, either silicon or germanium.

Germanium and Silicon contain semi-conductive material property and chemically pure. They are classified as intrinsic semiconductors. We can control their conductivity by controlling the amount of impurities added to this intrinsic semiconductor material. Free electrons or holes can be produced correspondingly by adding numerous impurities called donors or acceptors to this intrinsic material. This process of adding donor or acceptor atoms to semiconductor atoms (the order of 1 impurity atom per 10 million (or more) atoms of the semiconductor) is known as Doping. Because of not having purity, these donor and acceptor atoms of doped silicon are together known as "impurities". After doping these silicon material with a adequate number of impurities, it can be converted into an N-type or P-type semi-conductor material.

Most widely used basic semiconductor material is the silicon. It contains four valence electrons in its outermost shell. It shares this electron with its neighboring silicon atoms to form full orbital's of eight electrons. The bond structure between the two silicon atoms is such that each atom shares one electron with its neighbor making the bond very firm.

Very little free electrons are available to move around the silicon crystal. As a result, crystals of pure silicon (or germanium) are good insulators, or at the very least very high value resistors.

Silicon atoms are arranged in a definite regular shape making them a crystalline solid structure. A crystal of pure silica (silicon dioxide or glass) is usually called an intrinsic crystal (it has no impurities) and hence has no free electrons.

Just connecting a silicon crystal to a battery supply is not enough to extract an electric current from it. To do that we require to build a "positive" and a "negative" pole within the silicon permitting electrons and therefore electric current to flow out of the silicon. These poles are built by doping the silicon with certain impurities.

Structure of A Silicon Atom

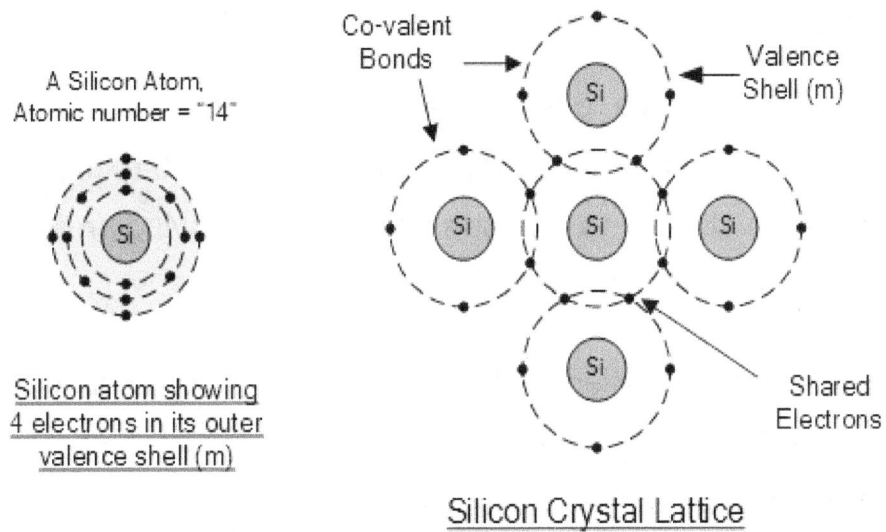

N-type Semiconductor

For conducting electricity by silicon crystal, we require to introduce an impurity atom like Arsenic, Antimony or Phosphorus into the crystalline structure creating it extrinsic (impurities are added). These atoms contain five outer electrons in their outermost orbital to share with neighboring atoms and are generally known as "Pentavalent" impurities.

This permits four out of the five orbital electrons to bond with its neighboring silicon atoms leaving one "free electron" to become mobile when an electrical voltage is applied (electron flow). As each impurity atom "donates" one electron, pentavalent atoms are usually called "donors".

Antimony (symbol Sb) as well as Phosphorus (symbol P), are often used as a pentavalent additive to silicon. Antimony has 51 electrons arranged in five shells around its nucleus with the outermost orbital contains five electrons. The resulting semiconductor basics material contains an excess of current-carrying electrons, each with a negative charge, and is therefore referred to as an N-type material with the electrons are known as "Majority Carriers" while the resulting holes are known as "Minority Carriers".

Stimulating by an external power source, electrons are released from the silicon atoms and quickly replaced by the free electrons available from the doped Antimony atoms. But this action still leaves an extra free electron moving around the doped crystal creating it negatively charged.

Then a semiconductor material is classified as N-type when its donor density is greater than its acceptor density, in other words, it has more electrons than holes thereby creating a negative pole as shown.

Antimony Atom and Doping

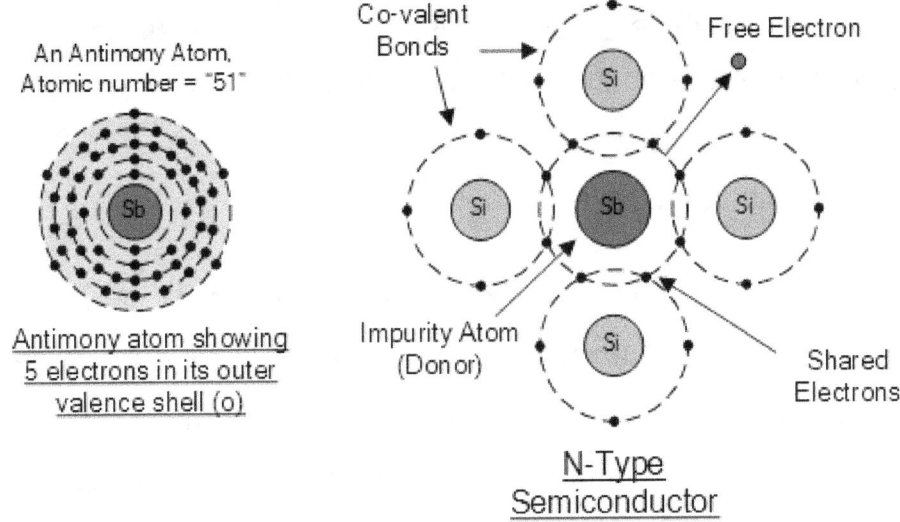

P-Type Semiconductor

The fourth closed bond cannot be formed if a "Trivalent" (3-electron) impurity is introduced into the crystalline structure, such as Aluminum, Boron, or Indium, which only have three valence electrons accessible in their outermost orbital. As a result, a complete connection is not feasible, resulting in an abundance of positively charged carriers known as holes in the crystal structure where electrons are essentially missing, giving the semiconductor material an abundance of positively charged carriers known as holes.

Because of having a hole in the silicon crystal, a neighboring electron is attracted to it and will try to travel into the hole to fill it. However, the electron filling the hole leaves another hole behind it as it travels. As a result, this attracts another electron which in turn creates another hole behind it, and therefore giving the presence that the holes are travelling as a positive charge through the crystal structure (conventional current flow).

Because to the displacement of holes, there are less electrons in the silicon, causing the entire doped crystal to become a positive pole. Trivalent impurities are known as "Acceptors" because they are constantly "accepting" extra or free electrons because each impurity atom causes a hole.

Boron (symbol B) is a frequent trivalent additive because it contains only five electrons distributed in three shells surrounding its nucleus, with only three electrons in the outermost orbital. Boron atom doping enables conduction to be dominated by positive charge carriers, resulting in a P-type material, with positive holes referred to as "Majority Carriers" and free electrons as "Minority Carriers."

When the acceptor density exceeds the donor density, a semiconductor basic material is classified as P-type. A P-type semiconductor has more holes than electrons as a result.

Boron Atom and Doping

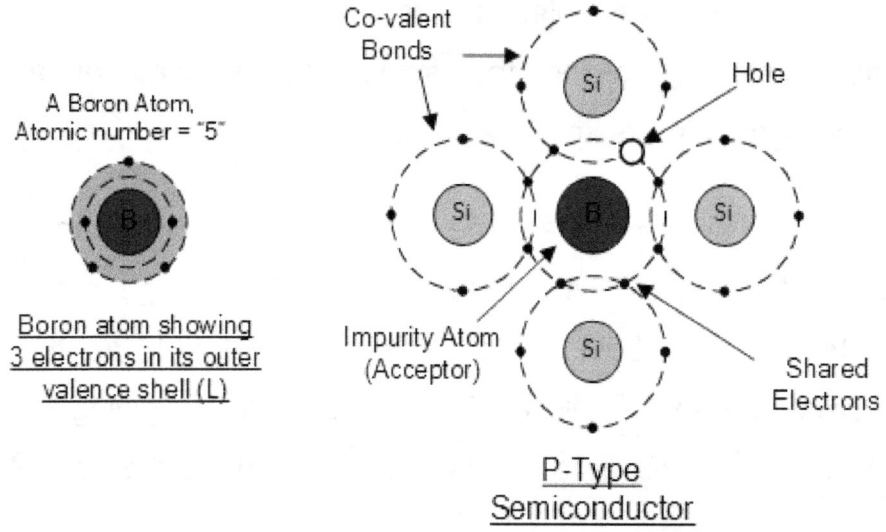

N-type (Doping by Antimony)

Materials added with Pentavalent impurity atoms (Donors) and accompanied by "electron" movement are known as N-type Semiconductors.

Following things happened in N-type semiconductors:

1. Positively charged donors.

2. Huge number of free electrons.

3. Very few holes in comparison to free electrons.

4. Doping provides:

> Donors with positively charged.
> Free electrons with negatively charged.

5. Energy supply provides:

> negatively charged free electrons.
> positively charged holes.

P-type (Doping by Boron)

Materials added with Trivalent impurity atoms (Acceptors) and accompanied by "hole" movement are known as P-type Semiconductors.

Following things happened in this type of materials:

1. Negatively charged acceptors.

2. Huge number of holes.

3. Very few free electrons compare to the number of holes.

4. Doping provides:

> ➤ Acceptors with negatively charged.
> ➤ Holes with positively charged.

5. Energy supply provides:

> ➤ positively charged holes.
> ➤ negatively charged free electrons.

Overall, both P and N-types are electrically neutral on their own.

Antimony and boron are two doping agents that are more commonly available than other materials. They are also known as "metalloids". There are a number of other chemical elements that share the same three- or five-electron outer-shell configuration, which makes them good choices for doping materials in the periodic table. Other elements can also be used to dope a base material of either silicon or germanium to produce different types of basic semiconductor materials for use in electronic semiconductor components, microprocessors, and solar cells. The following semiconductor materials are available.

Periodic Table of Semiconductors

Elements Group 13	Elements Group 14	Elements Group 15
3-Electrons in Outer Shell (Positively Charged)	4-Electrons in Outer Shell (Neutrally Charged)	5-Electrons in Outer Shell (Negatively Charged)
(5)	(6)	
Boron (B)	Carbon (C)	
(13)	(14)	(15)

Aluminium (Al) (31)	Silicon (Si) (32)	Phosphorus (P) (33)
Gallium (Ga)	Germanium (Ge)	Arsenic (As) (51)
		Antimony (Sb)

CHAPTER-2: THEORY OF PN JUNCTION

Theory of PN Junction

Newly doped N-type and P-type semiconductor materials are electrically neutral. After adding these two semiconductor materials together they uniquely integrate with each and form a junction which is called "**PN Junction** ". In this case, a big inclination is formed between both sides of the junction. Some free electrons from the donor impurity atoms start to move to the newly formed junction for filling up the holes in the P-type material creating negative ions. Therefore, they leave behind positively charged donor ions (N_D) on the negative side and now the holes from the acceptor impurity travel across the junction in the reverse direction into the region where there are lots of free electrons.

In this case, P-type's charge density along the junction is filled with negatively charged acceptor ions (N_A), and N-type's charge density along the junction becomes positive. This charge relocation of electrons and holes across the PN junction is called diffusion. P and N layers' width depends on how heavily each side is doped with acceptor density N_A, and donor density N_D, individually.

This progression remains to and fro until quantities of electrons which have overlapped the junction have sufficient electrical charge to prevent additional charge carriers from overpassing the junction. Ultimately a state of equilibrium (electrically neutral situation) will occur creating a "potential barrier" zone around the area of the junction as the donor atoms prevent the holes and the acceptor atoms prevent the electrons.

As no free charge carriers can rest in a position where there is a potential barrier, the regions on either sides of the junction now become completely depleted without any free carriers in comparison to the N and P type materials further away from the junction. This area around the **PN Junction** which is known as Depletion Layer.

The PN junction

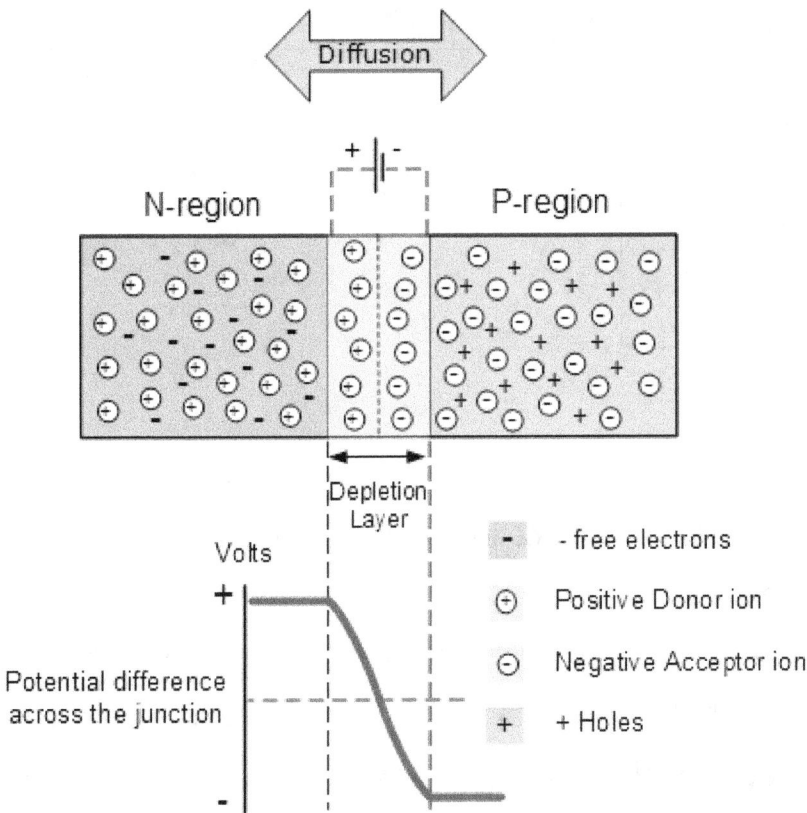

Total charge on both side of a *PN Junction* must be equal and reverse to keep a neutral charge condition around the junction. If the depletion layer region has a distance D, it must be entered into the silicon by a distance of Dp for the positive side, and a distance of Dn for the negative side giving a relationship between the two of: $Dp*N_A = Dn*N_D$ in order to maintain charge neutrality also known as equilibrium.

PN Junction Distance

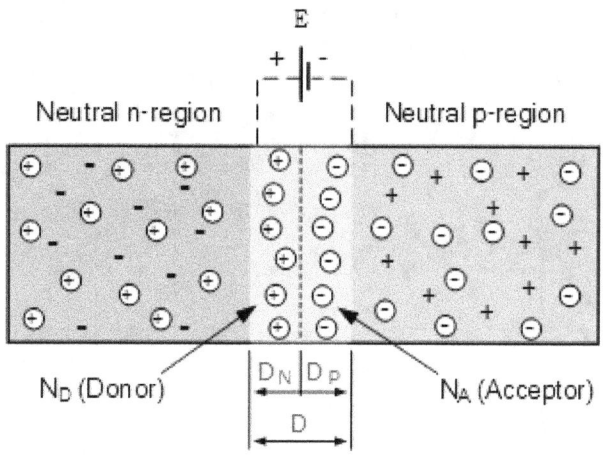

N-type material became positive with respect to the P-type because of losing electrons by N-type material and losing holes by the P-type. Then the existence of impurity ions on both sides of the junction cause an electric field to be established across this region with the N-side at a positive voltage relative to the P-side. A free charge needs additional energy to overcome the barrier that now exists for it to be able to cross the depletion region junction.

This electric field developed by the diffusion process has developed a "built-in potential difference" across the junction with an open-circuit (zero bias) potential of:

$$E_o = V_T \ln\left(\frac{N_D \cdot N_A}{n_i^2}\right)$$

Here, E_o is the zero bias junction voltage, V_T the thermal voltage of 26mV at room temperature, N_D and N_A are the impurity concentrations and n_i is the intrinsic concentration.

An appropriate positive voltage (forward bias) applied between the two ends of the PN junction can supply the free electrons and holes with additional energy.

The external voltage needed to overcome this potential barrier that now exists is very much reliant on the type of semiconductor material used and its definite temperature.

Usually, at room temperature the voltage across the depletion layer for silicon is about 0.6 – 0.7 volts and for germanium is about 0.3 – 0.35 volts. This potential barrier will always remain even if the device is not plugged in with any external power source, as observed in diodes.

As this built-in potential across the junction withstands both the flow of holes and electrons across the junction, it is known as potential barrier. In practice, a **PN junction** is formed within a single crystal of material rather than just simply connecting or combining together two distinct portions.

The result of this process is that the PN junction has current-voltage (IV or I-V) rectifier characteristics. Electrical contacts are fused on either side of the semiconductor to establish an electrical connection to an external circuit. The resulting electronics are often referred to as PN junction diodes or simply signal diodes.

Then we observed here that a PN junction can be made by joining or diffusing together different doped semiconductor materials to create an electronic device called a diode. can be used as the basic semiconductor structure of the rectifier, all kinds of transistors, LEDs, solar power. cells, and many other semiconductor devices.

CHAPTER-3: PN JUNCTION DIODE

PN Junction Diode

PN junction diodes formed when fusing a p-type semiconductor with an n-type semiconductor induces a potential barrier voltage across the diode junction, resulting in the junction being in equilibrium.

However, if we make an electrical connection to the ends of the N-type and P-type materials and then connect them to a battery source, an additional source of energy will now exist to overcome the potential barrier.

The addition of this extra energy causes electrons to freely pass through the depletion region from side to side. The operation of the PN junction with respect to the width of the potential barrier produces a disproportionately conductive double-ended device, known as a PN junction diode.

A PN Junction Diode is one of the simplest semiconductor devices around, and which has the electrical characteristic of passing current through itself in one direction only. However, unlike a resistor, a diode does not behave linearly with respect to the applied voltage. Instead it has an exponential current-voltage (IV) relationship and therefore we cannot describe its operation by simply using an equation such as Ohm's law.

If a suitable positive voltage (forward bias) is applied between the two ends of the PN junction, it can supply free electrons and holes with the extra energy they require to cross the junction as the width of the depletion layer around the PN junction is decreased.

By applying a negative voltage (reverse bias) results in the free charges being pulled away from the junction resulting in the depletion layer width being increased. This has the effect of increasing or decreasing the effective resistance of the junction itself allowing or blocking current to flow through the PN junction diodes.

Then the attenuation layer will widen as reverse voltage application is increasing and narrow as forward voltage application will increase. This is due to the difference in electrical characteristics on both sides of the PN junction, resulting in physical changes. One of the results produces rectification as seen in the static IV (current-voltage) characteristic of the PN junction diodes. Rectification is represented by asymmetric current when the polarity of bias voltage is changed as shown below.

Symbol and Static I-V Characteristics of Junction Diode

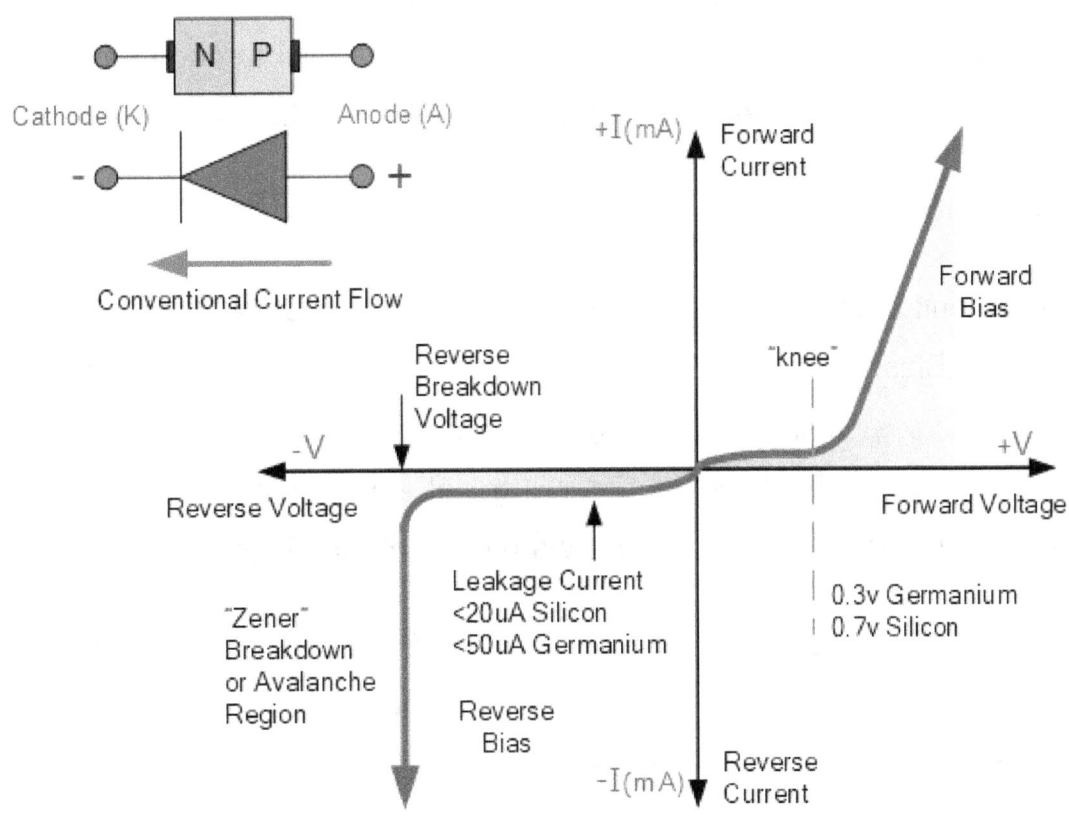

But before we can use the PN junction as an actual device or as a rectifier, we must first bias the junction, i.e. connect a voltage across it. On the voltage axis above, "Reverse bias" refers to an external voltage difference that increases the potential barrier. An external voltage that lowers the potential barrier is said to operate in the "Front Differentiation" direction.

There are two operating regions and three possible "bias" conditions for a standard junction diode, and these are:

1. **Zero Bias**: No external voltage potential is applied to the PN junction diode.
2. **Reverse Bias:** The voltage potential is connected negative, (-ve) to the P-type material and positive, (+ve) to the N-type material through the diode which has the effect of increasing the width of PN junction diode.
3. **Forward Bias**: The voltage potential is connected positive, (+ve) to the P-type material and negative, (-ve) to the N-type material through the diode which has the effect of decreasing the width of PN junction diode.

Zero Biased Junction Diode

No external potential energy is applied to the PN junction during connecting a diode in a Zero Bias condition. Therefore, if the diodes terminals are shorted together, a few holes (majority carriers) in the P-type material with adequate energy to overcome the potential barrier will travel across the junction against this barrier potential. This is called the "Forward Current" and is referenced as I_F

Similarly, holes generated in the N-type material (minority carriers), find this situation favorable and travel across the junction in the reverse direction. This is called the "Reverse Current" and is referenced as I_R. This relocation of electrons and holes back and forth across the PN junction is called diffusion, as shown below.

Zero Biased PN Junction Diode

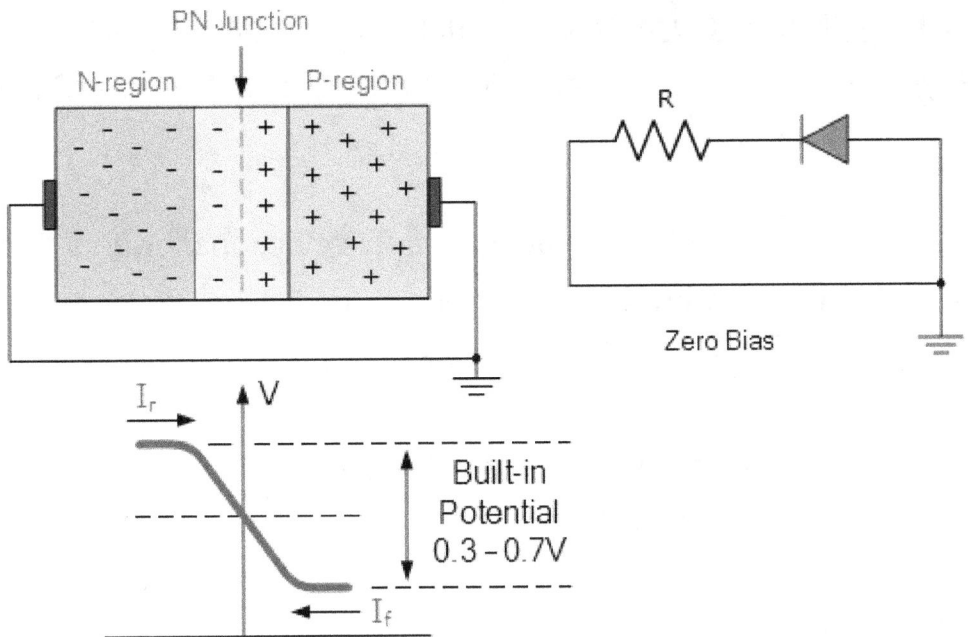

Existing potential barrier prevents the flow of additional majority carriers across the junction. Nevertheless, the potential barrier assists minority carriers (few free electrons in the P-region and few holes in the N-region) to drift across the junction.

Then an "Equilibrium" or balance will be formed when the majority carriers are equal and both moving in reverse directions, so that the net result is zero current flowing in the circuit. When this happens the junction is called in a state of "Dynamic Equilibrium ".

The minority carriers are frequently generated due to thermal energy so this state of equilibrium can be shattered by increasing the temperature of the PN junction causing an increase in the generation of minority carriers, thereby resulting in an increase in leakage current but an electric current cannot flow since no circuit has been connected to the PN junction.

Reverse Biased PN Junction Diode

A positive voltage is applied to the N-type material and a negative voltage is applied to the P-type material during connecting a diode in a Reverse Bias condition.

The positive voltage applied to the N-type material attracts electrons towards the positive electrode outside the junction, while the holes in the P-type end are also attracted outside the junction towards the negative electrode.

The net outcome is that the depletion layer develops broader due to a lack of electrons and holes and presents a high impedance path, almost an insulator and a high potential barrier is formed across the junction thus averting current from flowing through the semiconductor material.

Increase in the Depletion Layer due to Reverse Bias

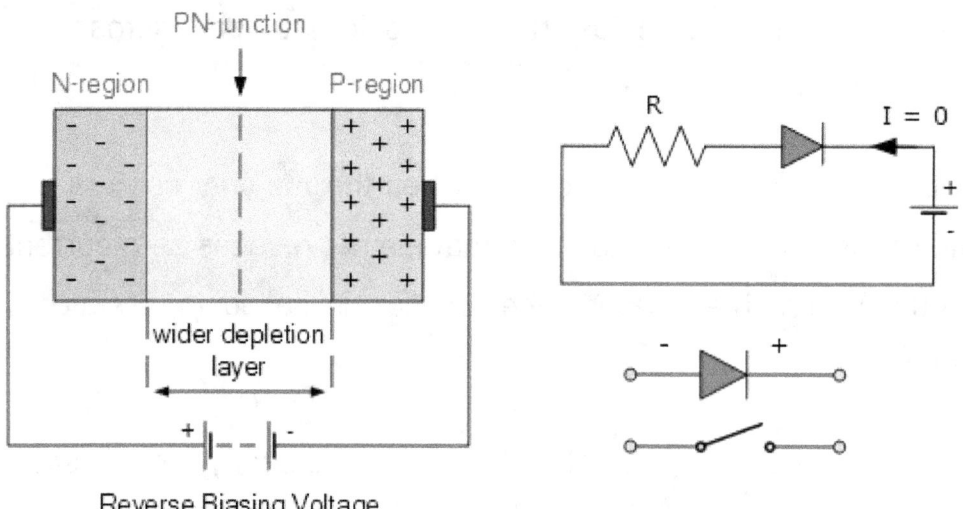

Reverse Biasing Voltage

This circumstance denotes a high resistance value to the PN junction and basically zero current flows through the junction diode with an upsurge in bias voltage. However, a slight reverse leakage current travel through the junction which is measured in micro-amperes, (µA).

If the reverse bias voltage Vr applied to the diode is very high, it will cause the diode's PN junction of the diode to overheat and fail due to the avalanche effect around the junction. This effect makes the diode shorted and causes the maximum circuit current movement, and this shown as a step descending incline in the reverse static characteristics curve below.

Reverse Characteristics Curve for a Junction Diode

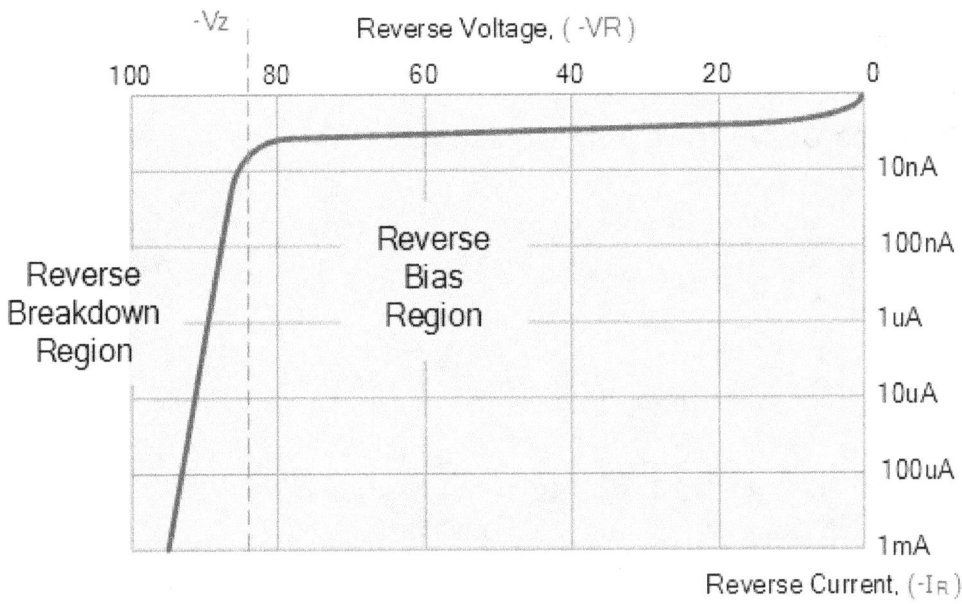

Occasionally this avalanche effect has concrete applications in voltage stabilizing circuits where a series limiting resistor is used with the diode to limit this reverse breakdown current to a predetermined maximum value thereby creating a fixed voltage output across the diode. These types of diodes are called Zener Diodes.

Forward Biased PN Junction Diode

A negative voltage is applied to the N-type material and a positive voltage is applied to the P-type material during connecting a Forward Bias. When this voltage is more than the potential barrier, approx. 0.7 volts for silicon and 0.3

volts for germanium, the potential barriers opposition will be overwhelmed and current will begin to flow.

As the negative voltage pushes electrons to the junction giving them the energy to cross over and associate with the holes being pushed in the reverse direction to the junction by the positive voltage. This consequence in a characteristics curve of zero current flowing up to this voltage point, known as "knee" on the static curves and then a high current flow through the diode with slight rise in the outside voltage as shown below.

Forward Characteristics Curve for a Junction Diode

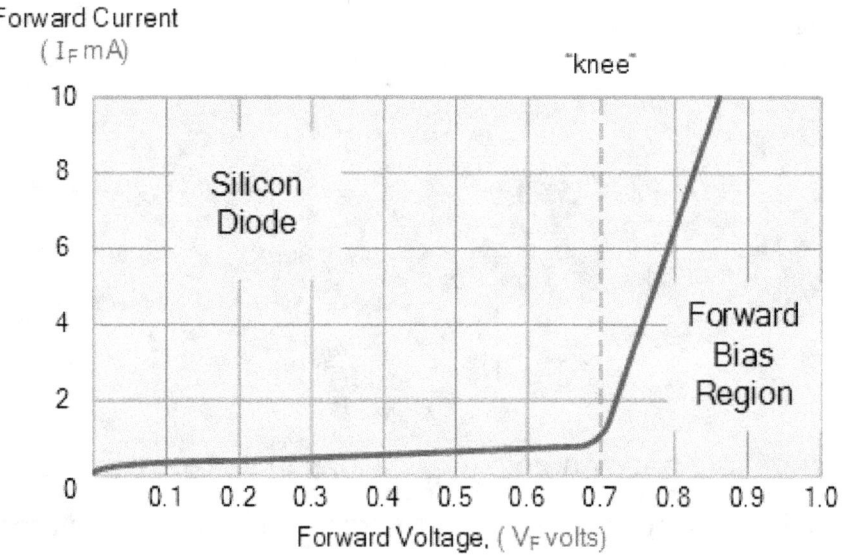

A forward biasing voltage's application on the junction diode causes the depletion layer becoming very tinny and thin which signifies a low impedance path through the junction thereby permitting high currents to move. The point at which this sudden upsurge in current takes place is denoted on the static I-V characteristics curve above as the "knee" point.

Reduction in the Depletion Layer due to Forward Bias

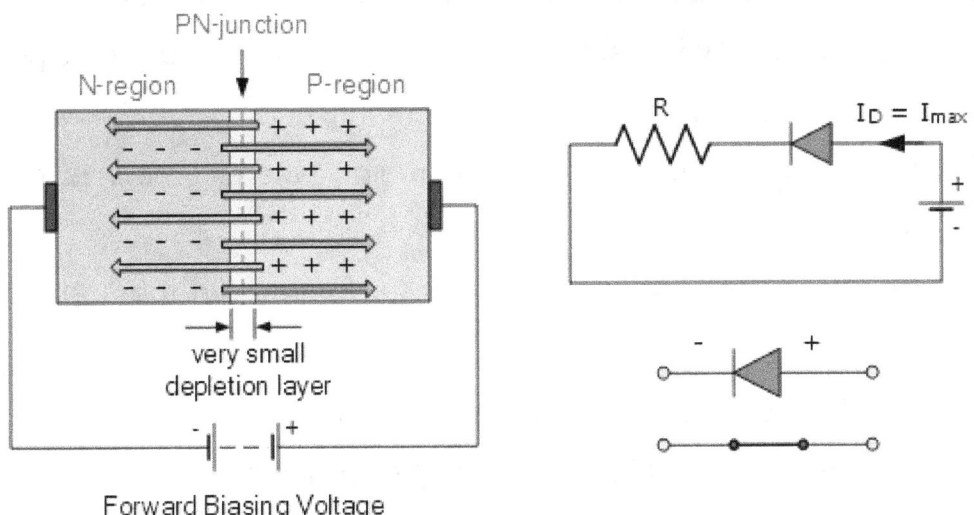

This circumstance signifies the low resistance path through the PN junction permitting huge currents to drift through the diode with only a little upsurge in bias voltage. The actual potential difference across the junction or diode is retained constant by the depletion layer's action at approximately 0.3v for germanium and approximately 0.7v for silicon junction diodes.

As the diode is able to conduct "infinite" current above this knee point, it causes a short circuit. Hence, resistors are used in series with the diode to limit the flow of current. Higher than its maximum forward current specification causes the device to dissipate more power in the form of heat than it was designed for causing rapid failure of the device.

Junction Diode at a glance:

A **J**unction Diode's PN junction region contains following significant features:

- There are two types of mobile charge carriers in semiconductors. They are: "Holes" and "Electrons".
- The holes are charged positively and the electrons are charged negatively.

- Doping a semiconductor with donor impurities like Antimony (N-type doping) comprises mobile charges which are predominantly electrons.
- Doping a semiconductor with acceptor impurities like Boron (P-type doping) comprises mobile charges which are primarily holes.
- The junction region is also called depletion region wherein there is no charge carriers.
- Depletion region has a physical thickness that fluctuates with the applied voltage.
- When a diode is Zero Biased no outside energy source is applied and a natural Potential Barrier is established in a depletion layer which is about 0.5 to 0.7v for silicon diodes and about 0.3 of a volt for germanium diodes.
- When a junction diode is Forward Biased the thickness of the depletion region decreases and the diode acts like a short circuit letting full circuit current to flow.
- When a junction diode is Reverse Biased the thickness of the depletion region rises and the diode acts like an open circuit blocking any current flow. In this case, only a very little leakage current will flow.
- Diodes are two terminal non-linear devices whose I-V characteristic are polarity dependent as rely on the polarity of the applied voltage, V_D the diode is either *Forward Biased*, $V_D > 0$ or *Reverse Biased*, $V_D < 0$. Either way we can model these current-voltage characteristics for both an ideal diode and for a real silicon diode.

Ideal and Real Characteristics of Junction Diode

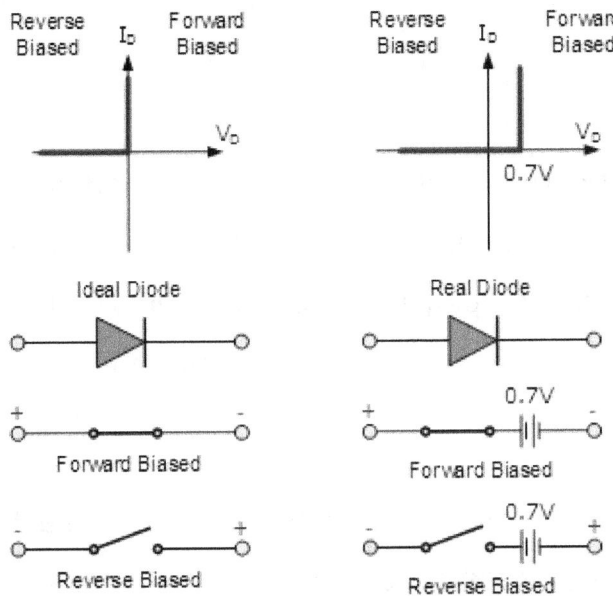

CHAPTER-4: SIGNAL DIODE

The Signal Diode

Small two-terminal signal diodes conduct current when forward biased and stop current flow when reverse biased. The semiconductor Signal Diode is a non-linear semiconductor device that is commonly employed in electronic circuits that include small currents or high frequencies, such as radio, television, and digital logic circuits.

When compared to larger Power Diodes, signal diodes such as the Point Contact Diode or the Glass Passivated Diode are physically quite small. A small signal diode's PN junction is normally enclosed in glass to preserve it, and it has a red or black band on one end of its body to help distinguish which end is the cathode terminal. The 1N4148 and its counterpart 1N914 signal diodes are the most commonly utilized of all the glass encased signal diodes.

When compared to rectifier diodes, small signal and switching diodes have much lower power and current ratings, around 150mA and 500mW maximum, but they can perform better in high frequency applications or clipping and switching applications that deal with short-duration pulse waveforms.

The characteristics of a signal point contact diode varies depending on whether it is made of germanium or silicon, and are as follows:

1. **Germanium Signal Diodes:** These have a low reverse resistance, resulting in a reduced forward volt drop across the junction (usually 0.2 to 0.3 volts), but a greater forward resistance due to their tiny junction area.
2. **Silicon Signal Diodes:** These have a high reverse resistance and a forward volt drop of 0.6 to 0.7 volts across the junction. They have

relatively low forward resistance, resulting in high forward current and reverse voltage peak values.

The electronic symbol for any type of diode is an arrow with a bar or line at one end, as shown below, along with the Steady State V-I Characteristics Curve.

V-I Characteristic Curve of Silicon Diode

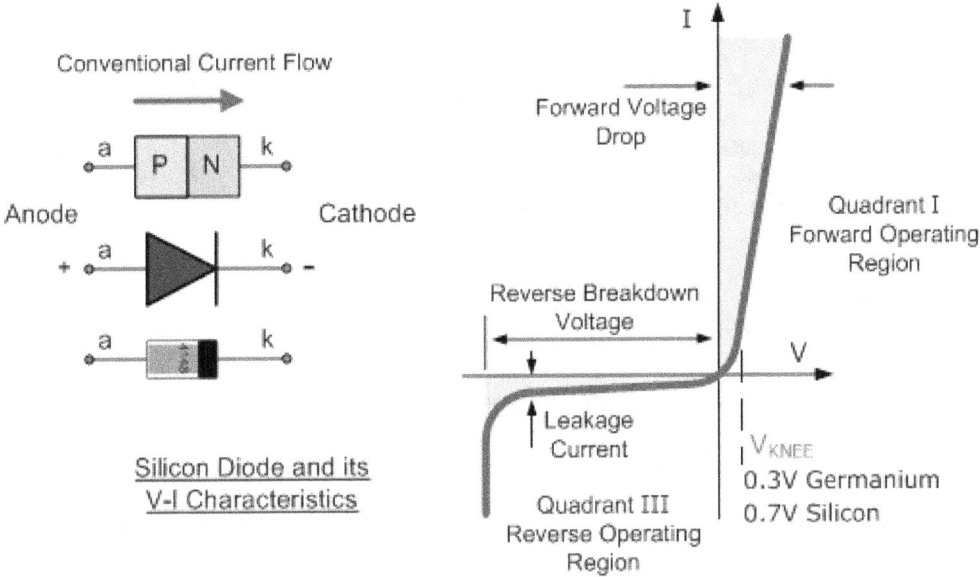

Silicon Diode and its V-I Characteristics

The arrow always points in the direction of the traditional current flow through the diode. That is, the diode conducts only when the positive power supply is connected to the anode terminal (a) and the negative power supply is connected to the cathode terminal (k), so the current flows in only one direction and is an electric one-way valve. It works like (forward).

However, when you connect an external power supply in the opposite direction, the diode blocks the current flowing through it and instead behaves like an open switch (reverse bias state) as shown below.

Forward and Reversed Biased Diode

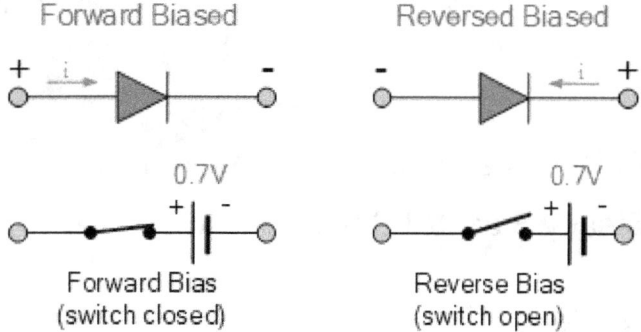

The ideal small signal diode passes current in one direction (forward conduction) and cut off current in the other direction (reverse blocking). Signal diodes are used in a variety of applications. Signal diodes are used as a rectifier, current limiter, voltage limiter, or switch for waveform shaping circuits.

Signal Diode Parameters

Signal diodes are manufactured in a wide range of voltage and current ratings and care should be taken when choosing a diode for a given application. There is a wide range of static characteristics associated with the modest signal diode.

1. Maximum Forward Current

The maximum forward current (IF(max)), as the name suggests, is the maximum forward current allowed to flow through the device. When the diode conducts under forward biased conditions, it has a very small "ON" resistance across the PN junction and so power is dissipated across this junction (Ohm's law) as heat. Then exceeding its value (IF(max)) will cause more heat to be generated across the junction and the diode will fail due to thermal overload, often with destructive consequences. When diodes operate around their maximum current rating, it is best to provide additional cooling to dissipate the heat generated by the diodes. For example, our small 1N4148 signal diode has a maximum current rating of about 150mA with a power dissipation of 500mW at 25°C. Then a resistor must be used in series with the diode to limit the forward current, (IF(max)) through it below this value.

2. Maximum Reverse Voltage

Maximum Reverse Voltage (VR (max)) or Peak Inverse voltage (PIV) or is the maximum allowable reverse operating voltage that can be applied across the diode without reverse failure or damage to the device. bag. Therefore, this rating is usually lower than the "avalanche fault" on the reverse bias characteristic curve. Typical values of VR (max) range from a few volts to thousands of volts and should be considered when replacing a diode.

Reverse peak voltage is an important parameter and is mainly used for rectification of diodes in ac rectifier circuits with reference to the magnitude of the voltage at which the sine waveform changes from positive to negative in each cycle.

3. Total Power Dissipation

The signal diodes have a total power dissipation, (PD(max)). This rating is the maximum possible power dissipation of the diode when forward biased (conductive). As current flows through the signal diode, the tendency of the PN junction is not perfect and creates some resistance to the current, resulting in dissipation (loss) of power in the diode as heat.

Since small signal diodes are nonlinear devices, the resistance of the PN junction is not constant, it is dynamic characteristic, so we cannot use Ohm's law to determine the power versus current and resistance or voltage and resistance as we can for resistors. Then to find the power that will be dissipated by the diode it is necessary to multiply the voltage drop at its terminals by the current passing through it: PD = V * I

4. Maximum Operating Temperature

Maximum operating temperature is actually related to diode junction temperature (TJ) and maximum power consumption. This is the maximum permissible temperature before the diode structure deteriorates and is expressed in degrees Celsius / watt (° C / W). This value is closely related to the maximum forward current of the device, so it will not exceed the junction temperature. However, the maximum passing current also depends on the ambient temperature at which the device operates, so the maximum passing current is typically given for two or more ambient temperature values, such as 25 ° C and 70 ° C.

Next, there are three main parameters to consider when choosing or replacing a signal diode. These are:

- The Reverse Voltage Rating
- The Forward Current Rating
- The Forward Power Dissipation Rating

Signal Diode Arrays

Diode arrays are very useful when space is limited or when switching signal diode pairs need to be matched. They generally consist of fast, low capacity silicon diodes such as the 1N4148 connected together in multiple diode packages called arrays for switching and clamping digital circuits. They are housed in a single line package (SIP) containing four or more internally connected diodes to provide a single isolated common cathode (CC) or common anode (CA) configuration as shown.

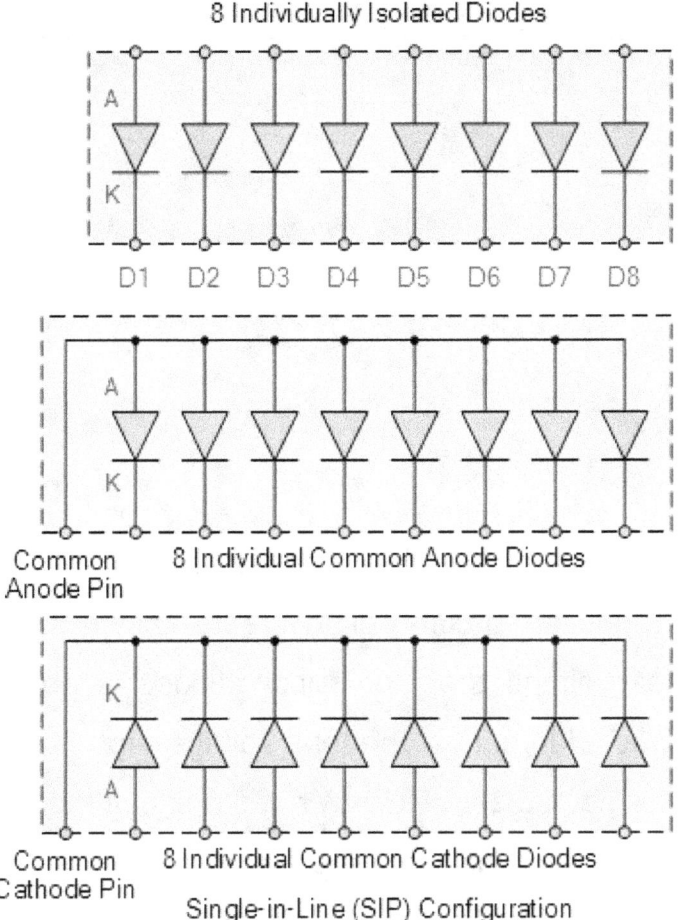

Single-in-Line (SIP) Configuration

Signal diode arrays can also be used in digital and computer circuits to protect high-speed data lines or other parallel input / output ports from electrostatic discharge (ESD) and transient voltages. By arranging two diodes in series across the power rail with the data line connected to the junction as shown, unwanted transients can be quickly dissipated and the signal diodes can be used in an 8-way array which is able to protect eight data lines in a single package.

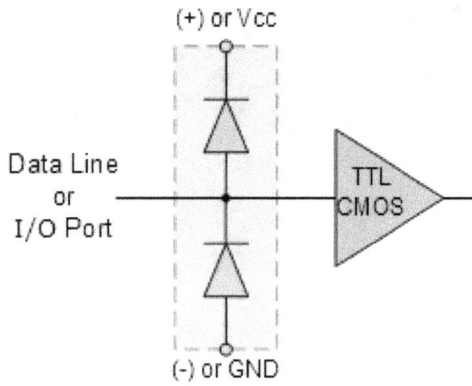

Protecting CPU Data Line

Signal diode arrays can also be used to connect diodes in series or in parallel to form voltage regulators or voltage reduction circuits, or to create known fixed reference voltages. The forward bias voltage drop for silicon diodes is about 0.7V, and if multiple diodes are connected in series, the total voltage drop is the sum of the individual voltage drops for each diode. However, if the signal diodes are connected in series, the current for each diode is the same and must not exceed the maximum forward current.

Signal Diodes' Series Connection

Another use for small signal diodes is to generate a regulated power supply. Diodes are connected in series to provide a constant DC voltage for the entire diode combination. The output voltage across the diode remains constant despite changes in the load current drawn by the series circuit or the DC supply voltage feeding the diode. Consider the following circuit.

Signal Diodes in Series

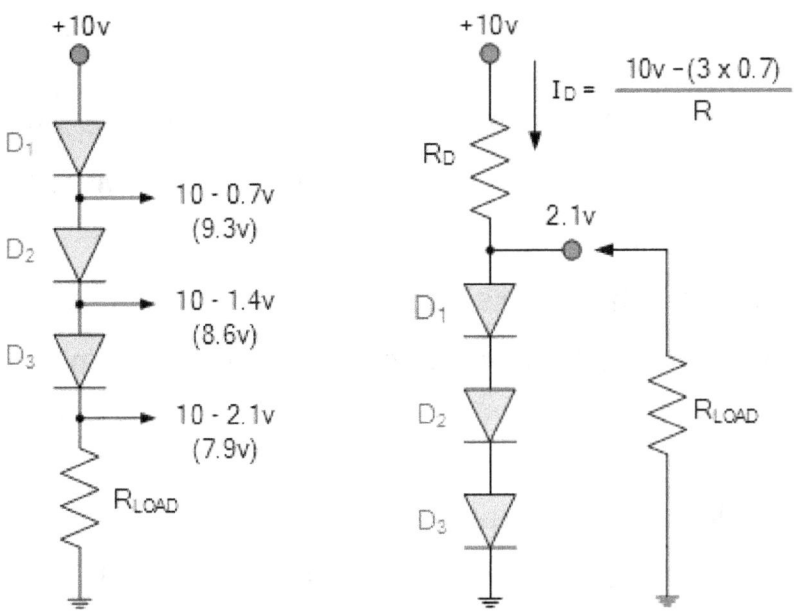

The forward voltage drop of a silicon diode is almost constant at about 0.7V, but the current flowing through the silicon diode fluctuates relatively large, so a forward bias signal diode can form a simple voltage regulation circuit. The individual voltage drop for each diode is subtracted from the supply voltage, leaving a specific potential across the load resistance. In the simple example above, this is given as 10V- (3 * 0.7V) = 7.9V. This is because each diode has contact resistance to the small signal current flowing through it, and the three signal diodes in series form three times that resistance, along with the load resistance R, forming voltage dividers across the power supply. Adding a diode in series will cause more voltage drop. You can also connect a diode connected in series in parallel with the load resistor to function as a voltage control loop. Here, the voltage applied to the load resistance is 3 * 0.7V = 2.1V. Of course, you can create the same constant voltage source with a single Zener diode. The resistor RD is used to prevent excessive current from flowing through the diode when the load is removed.

Freewheel Diodes

Signal diodes can also be used in a variety of clamping, protection, and waveform shaping circuits. The most common form of clamp diode circuit is to use a diode connected in parallel with a coil or inductive load to prevent damage to sensitive circuits. Suppression of voltage spikes and / or transients that occur when the load is suddenly switched "off". This type of diode is commonly known as a "freewheel diode", "flywheel diode", or simply a freewheel diode.

Freewheeling diodes are used to protect solid-state switches such as power transistors and MOSFETs from damage caused by reverse polarity protection and to protect them from highly inductive loads such as relay coils and motors. An example of the connection is shown below.

Use of the Freewheel Diode

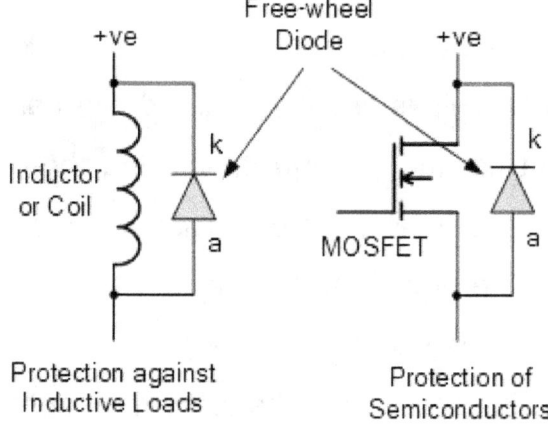

Modern fast switching, power semiconductor devices require fast switching diodes such as freewheeling diodes to protect them from inductive loads such as motor coils or relay windings. Every time the switching device above is turned "ON", the freewheel diode changes from a conducting state to a blocking state as it becomes reversed biased. However, when the device rapidly turns "OFF", the diode becomes forward biased and the collapse of the energy stored in the coil causes a current to flow through the freewheel diode. Without the protection of the freewheel diode high di/dt currents would occur causing a high voltage spike or transient to flow around the circuit possibly damaging the switching device. Previously, the operating speed of the semiconductor switching device, either transistor, MOSFET, IGBT or digital has been impaired by the addition of a freewheel diode across the inductive load with Schottky and Zener diodes being used instead in some applications. However, in recent years, freewheeling diodes have attracted attention mainly due to their improved reverse recovery characteristics and the use of ultrafast semiconductor materials that can operate at high switching frequencies. Other types of special diodes not included here are photodiodes, PIN diodes, tunnel diodes, and Schottky barrier diodes. Other types of semiconductor devices can be created by adding PN junctions to the basic two-layer diode structure. For example, a 3-layer semiconductor device can be a transistor, a 4-layer semiconductor device can be a thyristor or a silicon controlled rectifier, and a 5-layer device called a TRIAC can also be used.

CHAPTER-5: POWER DIODES AND RECTIFIERS

Power Diodes and Rectifiers

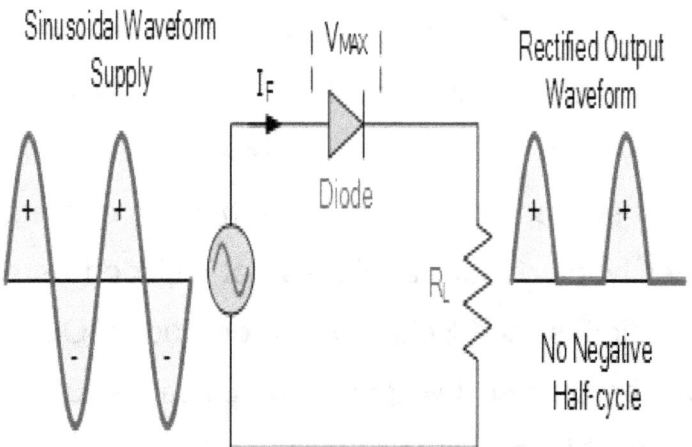

Power diodes are semiconductor PN junctions that can carry large currents at high voltage levels for use in rectifier circuits. This feature, and a common use for common diodes, is to convert alternating current (AC) to continuous current (DC). In other words, fix. Small signal diodes can be used in low power, low current (less than 1 amp) rectification and power applications. However, if a larger forward current or a higher reverse blocking voltage is required, the PN junction of the small signal diode will eventually overheat and be destroyed. Second, for high power applications, you should instead use larger and more robust power diodes. A power semiconductor diode, simply known as a power diode, has a much larger PN junction area than its small signal diode cousin, with a high forward current capacity of up to hundreds of amperes (KA) and maximum reverse direction. Brings blocking voltage up to thousands of volts (KV). Power diodes are not suitable for high frequency applications above 1MHz due to their large PN junctions, but special and expensive high frequency high current diodes are available. Schottky diodes are commonly used in high frequency and low voltage rectification applications due to their short reverse recovery time and low voltage drop in the forward biased state. Power diodes provide uncontrolled power rectification and are used in applications such as battery charging, DC power supplies, AC rectifiers and inverters. Due to their high current and voltage characteristics, they can also be used as freewheeling diodes and snubber networks. Power diodes are designed to have forward resistance that is a fraction of ohms, while reverse blocking resistance is in the mega ohm range. Some high quality power diodes are designed to be "stud mounted" on the heatsink, reducing thermal resistance to 0.1-1º C / watt. When an AC voltage is applied to a power diode, the diode conducts current during the positive half cycle, and during the negative half cycle, the diode does not conduct and cuts off the current flow. Second, conduction through the power diode is unidirectional because it only occurs during the positive half cycle. That's why it is Direct Current.

Power Diode Rectifier

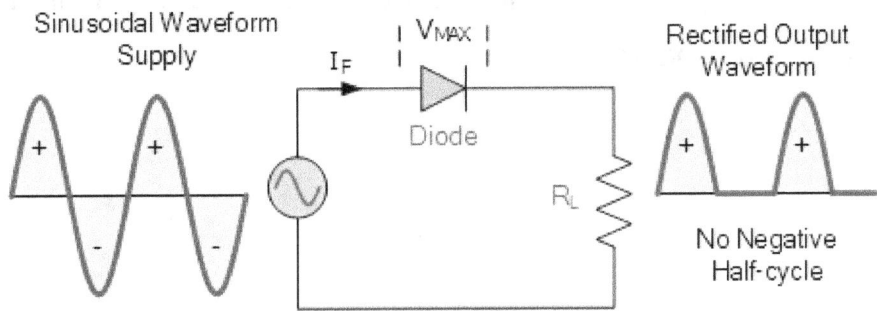

Power diodes can be used individually as described above or connected together to create a variety of rectifier circuits such as "half wave", "full wave" and "bridge rectifier". All types of commutators use uncontrolled, semi-controlled, uncontrolled commutators that use only power diodes, fully controlled commutators that use thyristors (SCRs), and semi-controlled commutators that are a combination of diodes. Or it can be classified as either full control. And thyristor.

The most commonly used single power diode in basic electrical device applications is the 1N400x series glass passivation general purpose rectifier diode. The standard continuous rectification forward current rating is approximately 1.0 amp, and the reverse blocking voltage rating is 50V for the 1N4001 and 1000V for the 1N4007. Here, the small 1N4007GP is the most popular for general line voltage rectification.

Half Wave Rectification

A rectifier is a circuit which converts the *Alternating Current* (AC) input power into a *Direct Current* (DC) output power. The input power supply may be either a single-phase or a multi-phase supply with the simplest of all the rectifier circuits being that of the **Half Wave Rectifier**.

The power diode in a half wave rectifier circuit passes just one half of each complete sine wave of the AC supply in order to convert it into a DC supply. Then this type of circuit is called a "half-wave" rectifier because it passes only half of the incoming AC power supply as shown below.

Half Wave Rectifier Circuit

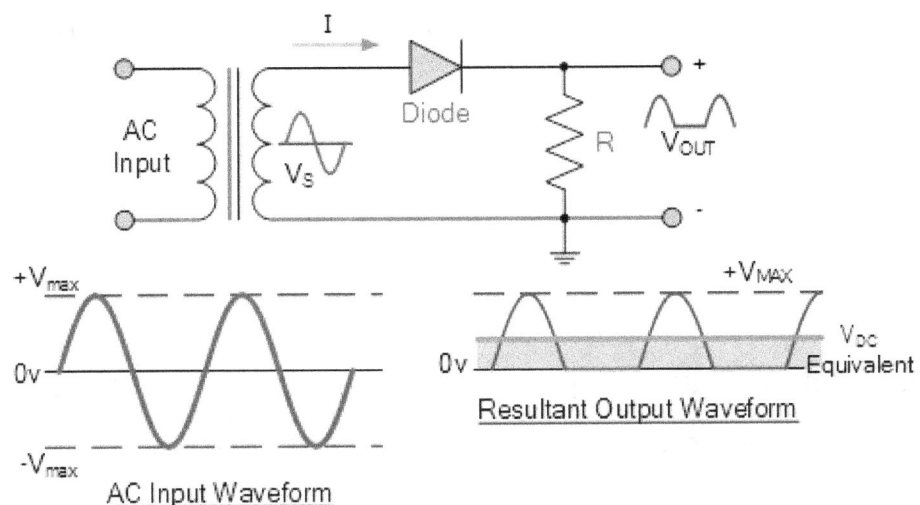

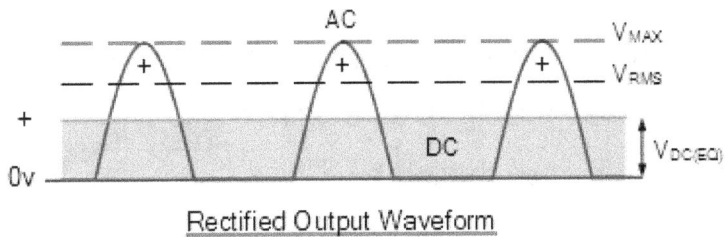

During each "positive" half cycle of the AC sine wave, the diode is forward biased as the anode is positive with respect to the cathode resulting in current flowing through the diode. Since the DC load is resistive (resistor, R), the current flowing in the load resistor is therefore proportional to the voltage (Ohm's Law), and the voltage across the load resistor will therefore be the same as the supply voltage, Vs (minus Vf), that is the "DC" voltage across the load is sinusoidal for the first half cycle only so Vout = Vs.

During each "negative" half cycle of the AC sinusoidal input waveform, the diode is reverse biased as the anode is negative with respect to the cathode. Therefore, NO current flows through the diode or circuit. Then in the negative half cycle of the supply, no current flows in the load resistor as no voltage appears across it so therefore, Vout = 0. The current on the DC side of the circuit flows in one direction only making the circuit Unidirectional. The load resistance from the diode receives positive half of the waveform, zero volt, positive half of the waveform, zero volt, etc., so this irregular voltage value is equal to the equivalent DC voltage of 0.318 * Vmax. Input sine wave, or 0.45 * Vrms sine wave input waveform. Next, calculate the equivalent DC voltage VDC across the load resistance as follows:

$$V_{d.c.} = \frac{V_{MAX}}{\pi} = 0.318 V_{MAX} = 0.45 V_{RMS}$$

Here, V_{MAX} is the maximum or peak voltage value of the AC sinusoidal supply, and V_{RMS} is the RMS (Root Mean Squared) value of the supply voltage.

Example of Power Diode:

Calculate the voltage drop V_{DC} and current I_{DC} flowing through a 100Ω resistor connected to a 240Vrms single-phase half-wave rectifier as described above. It also calculates the average DC power consumed by the load.

$$V_{MAX} = V_{RMS} \times 1.414, \quad \text{or} \quad V_{RMS} = V_{MAX} \times 0.7071$$

$$V_{DC} = 0.45 V_{RMS} = 0.45 \times 240 = 108 \text{ Volts}$$

or

$$V_{DC} = 0.318 V_{MAX} = 0.318 \times (240 \times 1.414) = 108 \text{ Volts}$$

$$I_{DC} = \frac{V_{DC}}{R} = \frac{108V}{100\Omega} = 1.08 \text{ Amps}$$

$$\text{Power} = I^2 R = 1.08^2 \times 100 = 116 \text{ Watts}$$

During rectification, the output DC voltage and current are both "ON" and "OFF" in each cycle. Since the voltage across the load resistor is present only for the positive half cycle (50% of the input waveform), this results in a low average DC value being fed to the load. Changing this rectified output waveform between the "ON" and "OFF" states will produce a waveform that exhibits a large amount of "ripple", which is an undesirable characteristic. Forms a DC ripple with a frequency equal to that of the AC power supply. Normally, when rectifying AC voltage, we want to produce a "smooth" and continuous DC voltage, without any voltage fluctuations or ripples. One way to do this is to connect a large value capacitor across the output voltage terminals in parallel with the load resistor as shown below. This type of capacitor is often referred to as a smoothing capacitor.

Half-wave Rectifier with Smoothing Capacitor

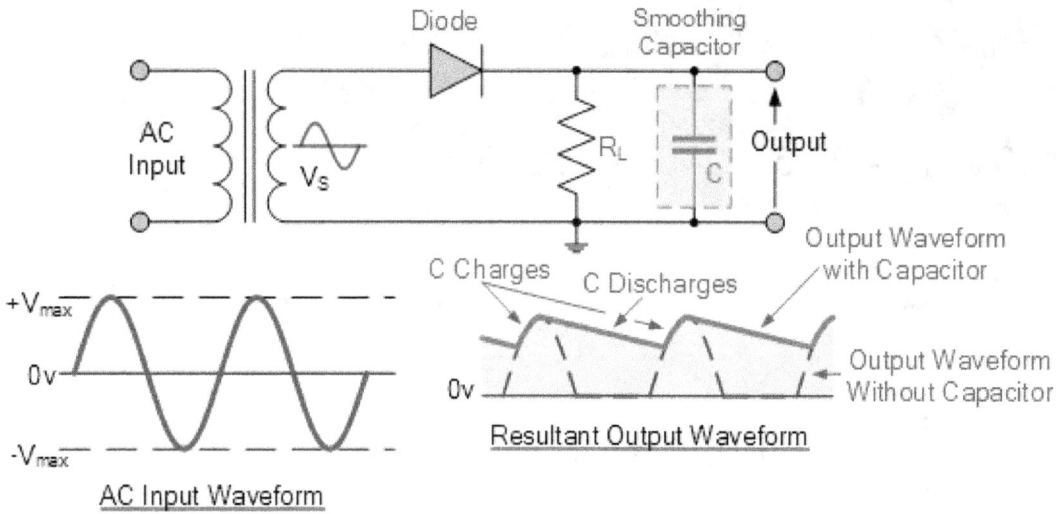

When a rectifier is used to supply a direct voltage (DC) source from an alternating current (AC) source, the amount of ripple voltage can be further reduced by using larger rated capacitors, but There are limitations on both the cost and size of smoothing capacitors. used. For a given capacitor value, a larger load current (lower load resistance) will discharge the capacitor faster (RC Time Constant) and thus increase the generated ripple. Then, for a single-phase half-wave rectifier circuit using power diodes, it is not very practical to try to reduce the ripple voltage by smoothing the capacitor. In this case, it is more convenient to use "Full Wave Rectification". In practice, half-wave rectifiers are used most often in low-power applications because of their major disadvantages. The output amplitude is less than the input amplitude, there is no output for the negative half cycle, so half the power is wasted and the output is pulsed to DC, resulting in overshoot. To overcome these disadvantages, several power diodes are connected together to create a full wave rectifier.

CHAPTER-6: FULL WAVE RECTIFIER

Full Wave Rectifier

Power diodes can be connected together to form a full-wave rectifier that converts an AC voltage to a pulsating DC voltage for use in the power supply.

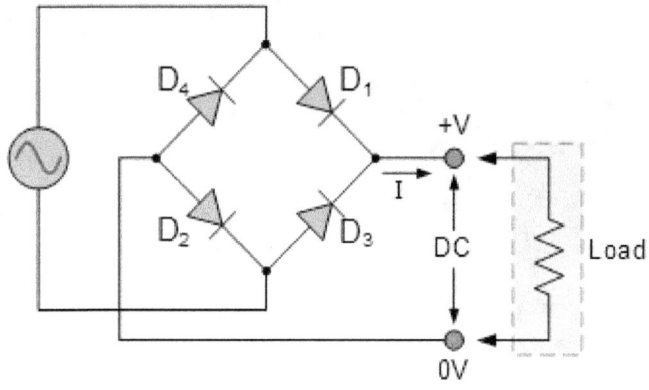

We know that connecting a smoothing capacitor to the load resistor can reduce voltage ripple or DC voltage fluctuations. This method is suitable for low power applications, but not for applications that require a "stable and smooth" DC supply voltage. One way to develop this is to use it every half cycle of the input voltage, not every half cycle. The circuit that makes this possible is called a full-wave rectifier. Like half-wave circuits, full-wave rectifier circuits produce an output voltage or current that has a constant or constant component. Full wave rectifiers have some fundamental advantages over their half wave rectifier counterparts. The average (DC) output voltage is higher than for half wave, the output of the full wave rectifier has much less ripple than that of the half wave rectifier producing a smoother output waveform. In a Full Wave Rectifier circuit two diodes are now used, one for each half of the cycle. A multiple winding transformer is used whose secondary winding is split equally into two halves with a common center tapped connection, (C). This configuration results in each diode conducting in turn when its anode terminal is positive with respect to the transformer center point C producing an output during both half cycles, twice that for the half wave rectifier so it is 100% efficient as shown below.

Full Wave Rectifier Circuit

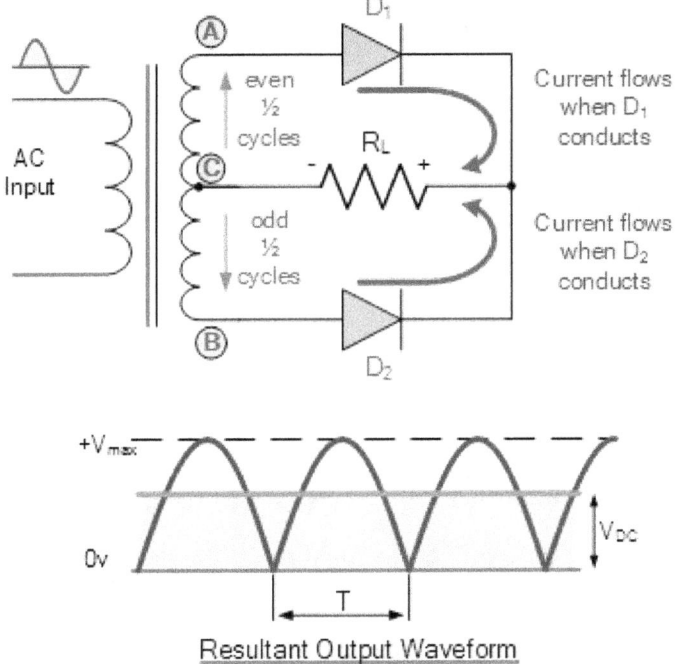

Resultant Output Waveform

A full-wave rectifier circuit consists of two power diodes connected to a load resistor (R_L), each taking it in turn to supply current to the load. When point A of the transformer is positive relative to point C, diode D1 conducts in the forward direction as indicated by the arrows.

When point B is positively charged (for a negative half cycle) relative to point C, diode D2 conducts in the forward direction and current through resistor R is in the same direction in both half cycles. Since the output voltage across the resistor R is the sum of the phasors of the two combined waveforms, this type of full-wave rectifier circuit is also known as a "bi-phase" circuit.

We can see this effect quite clearly if we run the circuit in a simulator with the smoothing capacitor removed.

Simulation Waveform

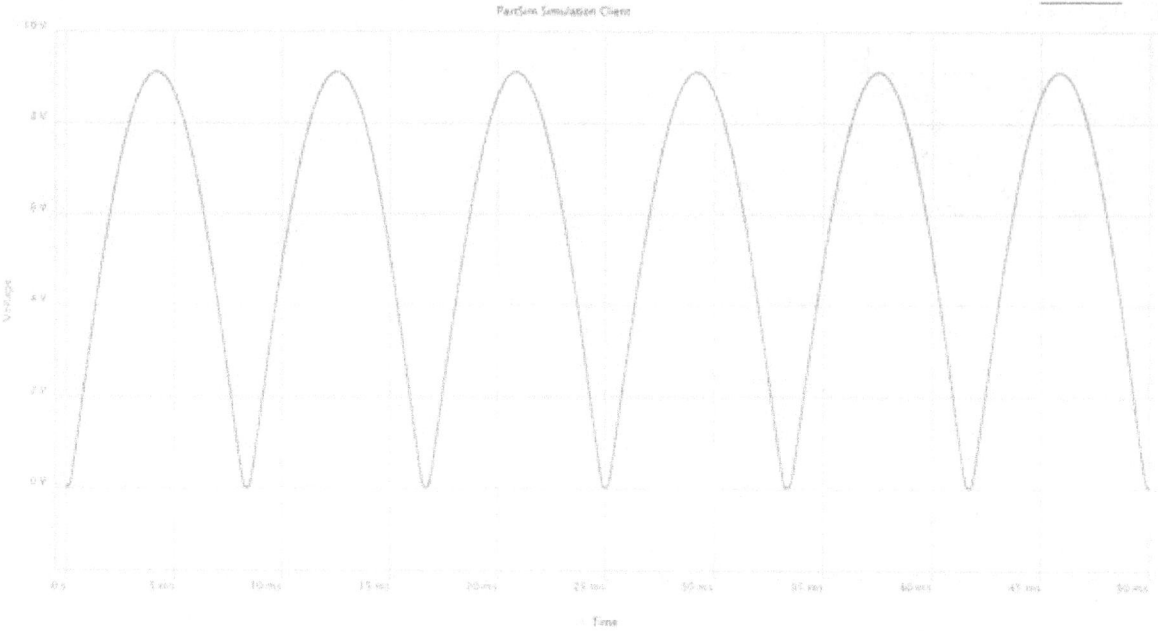

As the space between each half-wave developed by each diode is now filled by the other, the average DC output voltage across the load resistor is now twice that of a single half-wave rectifier circuit and is about 0.637 V_{max} of the peak voltage, assuming no losses.

$$V_{d.c.} = \frac{2V_{max}}{\pi} = 0.637 V_{max} = 0.9 V_{RMS}$$

Here, V_{max} is the maximum peak value in one half of the secondary winding and V_{RMS} is the RMS value as: $V_{RMS} = 0.7071 V_{max}$. The DC current is given as: $I_{DC} = V_{DC}/R$.

The peak voltage of the output waveform is the same as before for the half-wave rectifier provided that each half-winding of the transformer has the same value of RMS voltage. To achieve a different DC voltage output, different transformer ratios can be used.

The main disadvantage of this type of full-wave rectifier circuit is the need for a larger transformer for a given output power with two separate but identical secondary windings, which makes this type of full-wave rectifier circuit expensive compared to the equivalent "full-wave bridge rectifier" circuit.

The Full Wave Bridge Rectifier

Another type of circuit that produces an output waveform similar to the full-wave rectifier circuit above is a full-wave bridge rectifier. This type of single-phase rectifier uses four individual rectifier diodes connected in a closed-loop "bridge" configuration to produce the desired output.

The main advantage of this bridge circuit is that it does not require a special centering transformer, which reduces its size and cost. The single secondary coil is connected to one side of the diode bridge network and loaded to the other side as shown below.

The Diode Bridge Rectifier

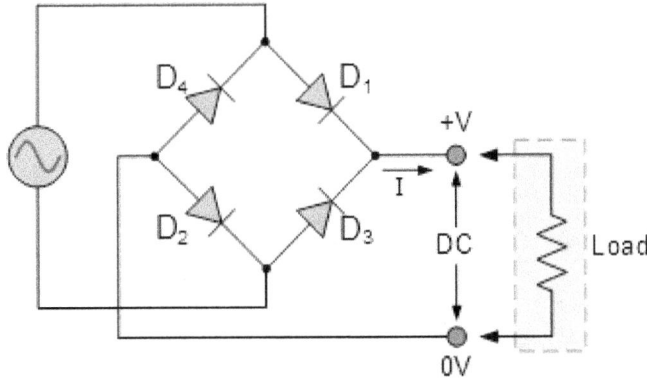

The four diodes labeled D1 to D4 are arranged as "series pair" with only two diodes carrying current in each half cycle. During the positive half cycle of the power supply, diodes D1 and D2 are in series while diodes D3 and D4 are reverse biased and current flows through the load as shown below.

The Positive Half-cycle

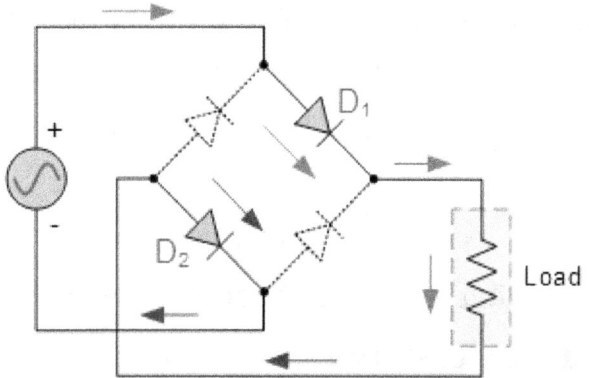

During the negative half cycle of the supply, diodes D3 and D4 are in series, but diodes D1 and D2 are off because they are reverse biased. The current through the load has the same direction as before.

The Negative Half-cycle

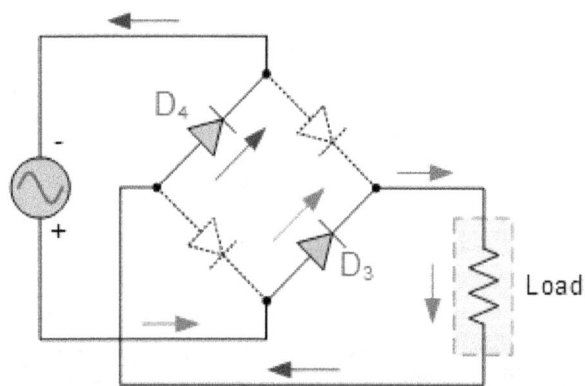

Since the current through the load is DC, the voltage developed across the load is also DC as with the previous two-diode rectifier, so the average DC voltage across the load is 0.637 V_{max}.

Typical Bridge Rectifier

In practice, however, in each half cycle current flows through two diodes instead of just one, so the magnitude of the output voltage is that the two voltages drop lower (2 * 0.7 = 1, 4 V) at the V_{max} input amplitude. The ripple frequency is now twice the supply frequency (e.g. 100Hz for a 50Hz supply or 120Hz for a 60Hz supply.)

While four individual power diodes can be used to create a full-wave bridge rectifier, pre-manufactured bridge rectifier components are available "on the shelf" in a wide range of other voltage and current sizes. each other can be soldered directly to the printed circuit board or connected by spade connectors.

A typical single-phase rectifier bridge has a cut-off angle. This cutoff indicates that the terminal closest to the corner is the positive or +ve output wire or terminal, with the opposite (diagonal) wire being the negative or output wire. The other two connection wires are for the AC input voltage of the transformer secondary winding.

The Smoothing Capacitor

We know that a half wave single phase rectifier produces an output wave every half cycle, and it is not practical to use this type of circuit to generate a continuous DC power supply. However, the full-wave bridge rectifier gives us a higher average continuous value (0.637 V_{max}) with less ripple superposition while the output waveform is twice the frequency of the input power supply frequency.

We can develop the average DC output of the rectifier while reducing the AC variation of the rectifier output by using smoothing capacitors to filter the output waveform. Smoothing capacitors or condensers connected in parallel with the load through the output of the full-wave bridge rectifier further increases the average DC output because the capacitor acts as a storage device as shown below.

Full-wave Rectifier with Smoothing Capacitor

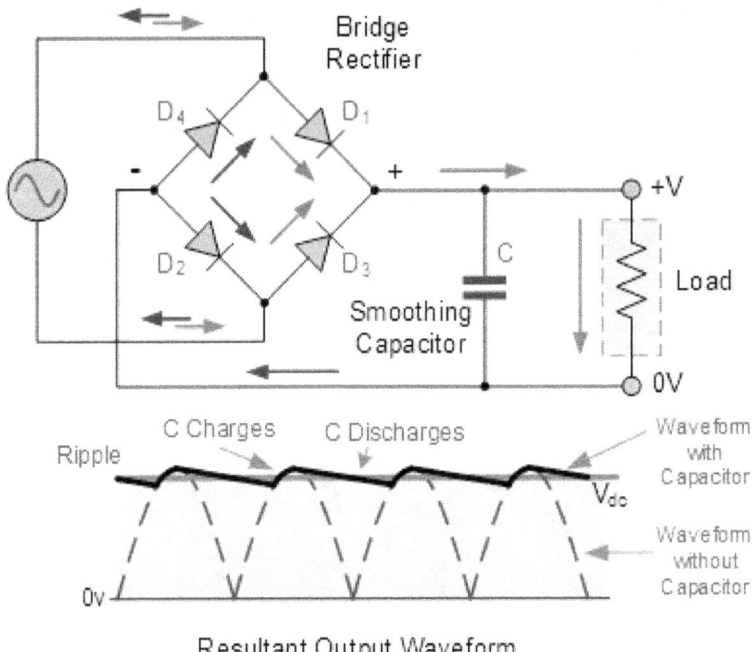

Resultant Output Waveform

The smoothing capacitor converts the rippled full-wave output of the rectifier into a smoother DC output voltage. Now, if we run the simulator circuit with different values of the installed smoothing capacitors, we can see the effect on the rectified output waveform as shown in the figure.

5uF Smoothing Capacitor

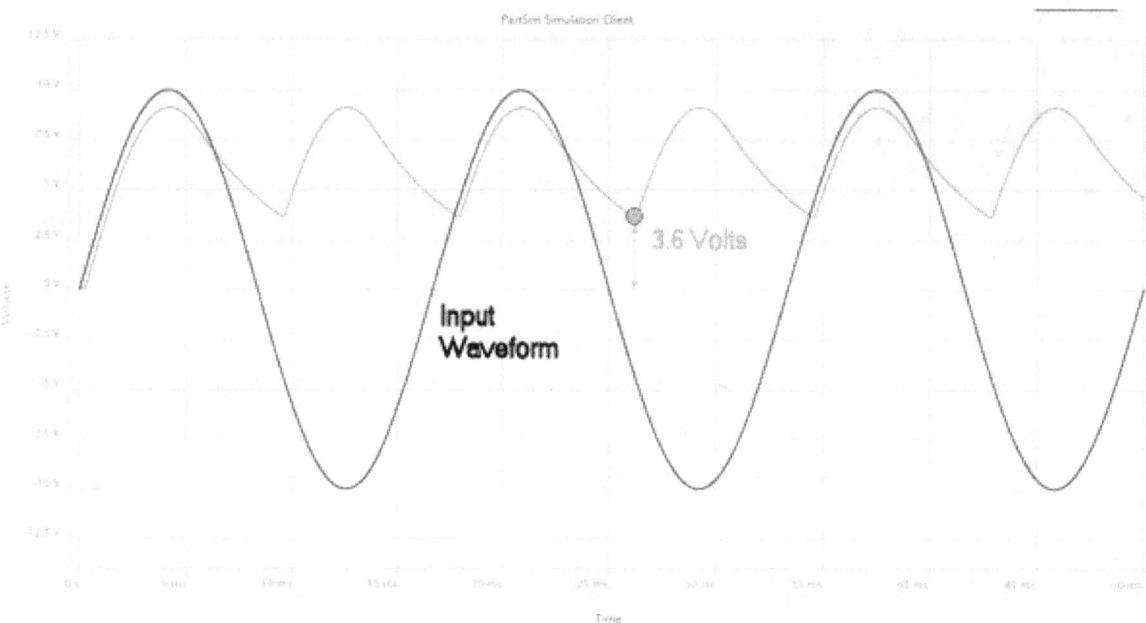

The plot having blue color on the waveform displays the result of using a 5.0uF smoothing capacitor across the rectifiers output. Previously the load voltage followed the rectified output waveform down to zero volts. Here the 5uF capacitor is charged to the peak voltage of the output DC pulse, but when it drops from its peak voltage back down to zero volts, the capacitor cannot discharge fast because of the RC time constant of the circuit.

This outcome in the capacitor discharging down to about 3.6 volts, in this example, maintaining the voltage across the load resistor until the capacitor recharges once again on the next positive slope of the DC pulse. In other words, the capacitor only has time to discharge briefly before the next DC pulse recharges it back up to the peak value.

In this way, the DC voltage applied to the load resistor drops only by a little amount. But we can develop this still by increasing the value of the smoothing capacitor as shown.

50uF Smoothing Capacitor

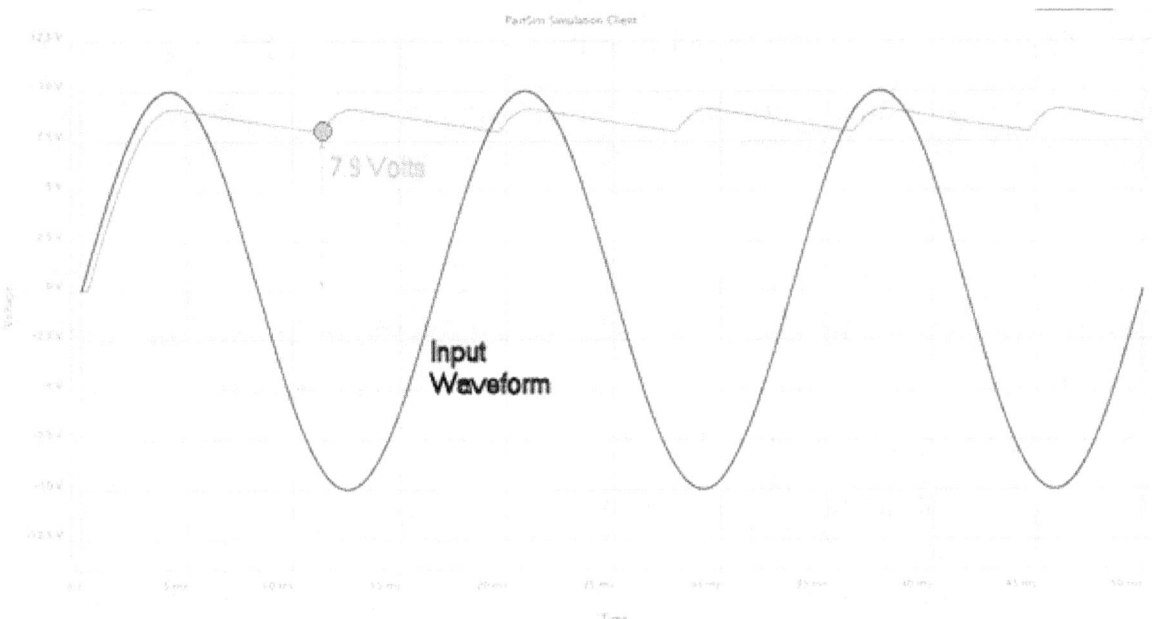

In this case, we have increased the value of the smoothing capacitor ten-fold from 5uF to 50uF which has reduced the ripple increasing the minimum discharge voltage from the previous 3.6 volts to 7.9 volts. Nevertheless, using the Simulator Circuit we have chosen a load of 1kΩ to achieve these values, but as the load impedance decreases the load current increases triggering the capacitor to discharge more quickly between charging pulses.

The effect of a providing a heavy load with a single smoothing capacitor can be reduced by the use of a larger capacitor which stores more energy and discharges less between charging pulses. Usually, for DC power supply circuits the smoothing capacitor is an Aluminum Electrolytic type that has a capacitance value of 100uF or more with repeated DC voltage pulses from the rectifier charging up the capacitor to peak voltage.

Nevertheless, there are two important parameters to consider when choosing a suitable smoothing capacitor and these are its *Working Voltage*, which must be greater than the no-load output value of the rectifier and its *Capacitance Value*, which regulates the amount of ripple that will appear overlaid on top of the DC voltage.

Too little a capacitance value and the capacitor has miniature effect on the output waveform. But if the smoothing capacitor is adequately large enough (parallel capacitors can be used) and the load current is not too large, the output voltage will be nearly as smooth as pure DC. As a general rule of thumb, we are looking to have a ripple voltage of less than 100mV peak to peak.

The maximum ripple voltage present for a Full Wave Rectifier circuit is not only regulated by the value of the smoothing capacitor but by the frequency and load current, and is calculated as:

Bridge Rectifier Ripple Voltage

$$V_{(ripple)} = \frac{I_{DC}}{2fC}, \text{ Volts}$$

Here, I is the DC load current in amps, f is the frequency of the ripple or twice the input frequency in Hertz, and C is the capacitance in Farads.

The main advantage of a full-wave bridge rectifier is that it has a lower AC ripple value for a particular load and has a smaller smoothing capacitor than an equivalent half-wave rectifier. Therefore, the fundamental frequency of the ripple voltage is twice the mains frequency (100 Hz), but in the case of a half-wave rectifier it is exactly the same as the mains frequency (50 Hz).

The magnitude of the ripple voltage overlaid on the DC supply voltage by the diode can be substantially eliminated by adding a significantly improved π (pi) filter to the output terminal of the bridge rectifier. This type of low pass filter usually consists of two smoothing capacitors of the same value and a choke or inductor across them, introducing a high impedance path to the AC ripple component.

Another more practical and cheaper alternative is to use off-the-shelf 3-terminal voltage regulator ICs such as: B. Positive output voltage or vice versa LM78xx ("xx" is output voltage), negative output voltage LM79xx (datasheet) that can reduce ripple by more than 70 dB while providing constant output current Power supply of 1 amp or more.

CHAPTER-7: THE ZENER DIODE

The Zener Diode

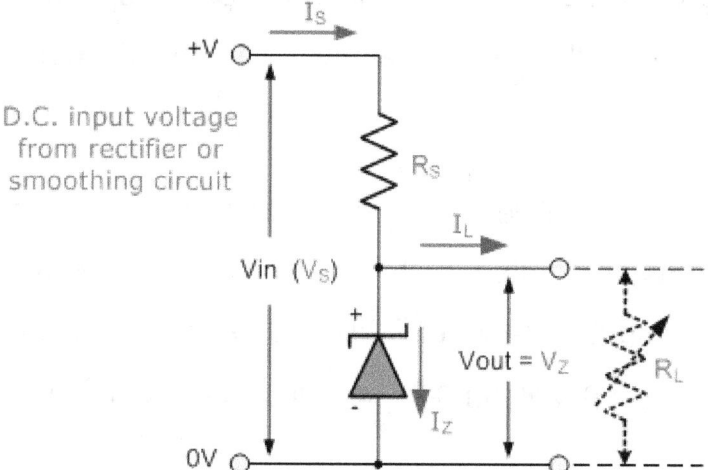

A semiconductor diode blocks current in the opposite direction, but will fail prematurely if the reverse voltage applied across the terminals becomes too high. However, Zener diodes or "breakdown diodes" as they are sometimes called, are essentially the same as standard PN junction diodes, but specially designed to have a specified low reverse breakdown voltage taking advantage of any reverse voltage is applied. to her. The Zener diode works like a regular general purpose diode made of PN silicon junction and when it is forward biased i.e. anode to its cathode it behaves like a normal signal diode carry rated current. However, unlike a conventional diode which blocks all current flowing through itself when reverse biased i.e. cathode becomes more positive than anode, as soon as reverse voltage reaches a predetermined value, Zener diode start to lead in the opposite direction. This is because when the reverse voltage applied across the Zener diode exceeds the rated voltage of the device, a process called Avalanche Breakdown occurs in the depletion layer of the semiconductor and a current begins to flow through the diode to limit it. this voltage rise. The current flowing through the Zener diodes increases suddenly to the maximum value of the circuit (usually limited by a series resistor) and when this reverse saturation is reached, this reverse saturation current remains fairly stable. over a wide range of reverse voltages. The voltage point at which the voltage across the zener diode becomes stable is known as the "zener voltage", (V_z) and for zener diodes this can range from less than one volt to several hundred volts. The point at which the zener voltage triggers current through the diode can be controlled very precisely (within a tolerance of 1%) in the doping step of the diode's semiconductor construction giving the diode a specific zener breakdown voltage. possible, (V_z) for example, 4.3V or 7.5V. This Zener breakdown voltage on the IV curve is almost a vertical line.

Zener Diode I-V Characteristics

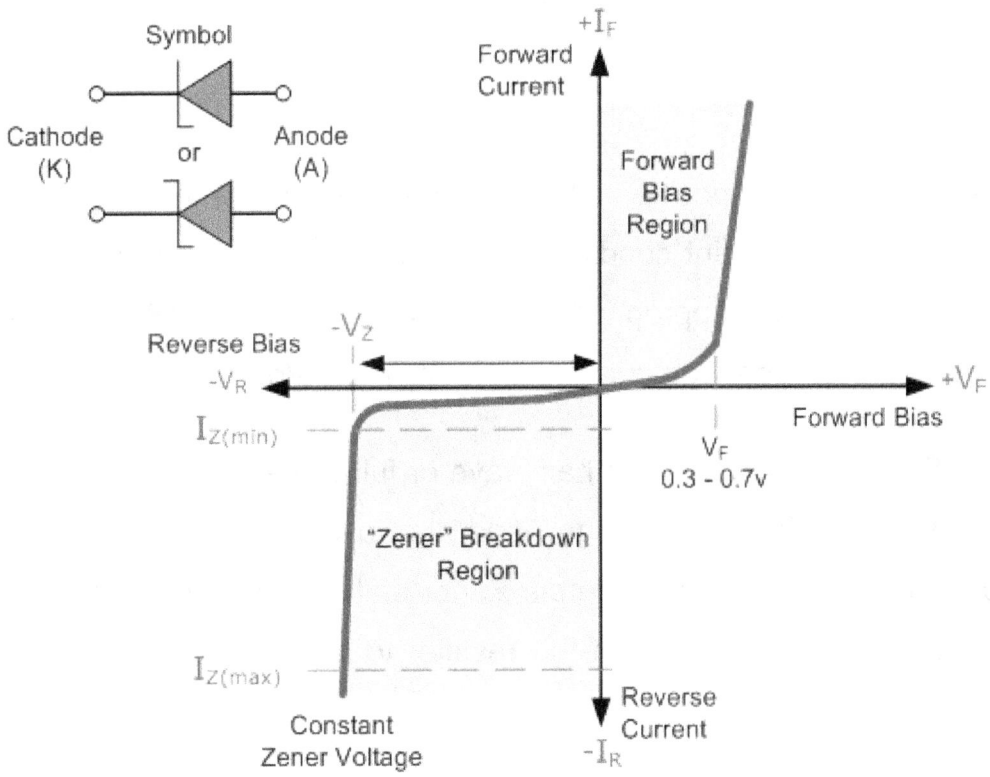

Zener diodes are used in "reverse bias" or reverse breakdown, i.e. the anode of the diodes connects to the negative supply. From the above IV characteristic, we can see that the Zener diode has a region in its reverse bias characteristic where a negative voltage is almost constant regardless of the amount of current flowing through the diode. This voltage is virtually constant even with large current variations provided that the current of the Zener diodes remains between the breakdown current I_Z (minimum) and the rated current I_Z maximum (maximum). The self-control of this Zener diode can be used well in regulating or stabilizing a voltage source against changes in supply or load. The fact that the voltage across the diode in the breakdown region remains virtually constant proves this is an important property of the Zener diode as it can be used in the simplest types of voltage regulation applications. The function of a voltage regulator is to provide a constant output voltage to a load connected in parallel despite ripples in the supply voltage or variation in load current. The Zener diode will continue to regulate its voltage until the current holding diode drops below the minimum value I_Z (min) in the reverse breakdown region.

The Zener Diode Regulator

Zener diodes can be used to generate stablized voltage outputs with less ripple under a variety of load and current conditions. By passing a small current from the voltage source through the appropriate current limiting resistor (RS) to the diode, the Zener diode draws enough current to maintain the Vout voltage drop.

We know that the DC output voltage of a half-wave or full-wave rectifier contains ripples superimposed on the DC voltage, and the average output voltage changes as the load value changes. A simple Zener stabilizer circuit, as shown below, can be connected to the output of the rectifier to produce a more stable output voltage.

Zener Diode Regulator

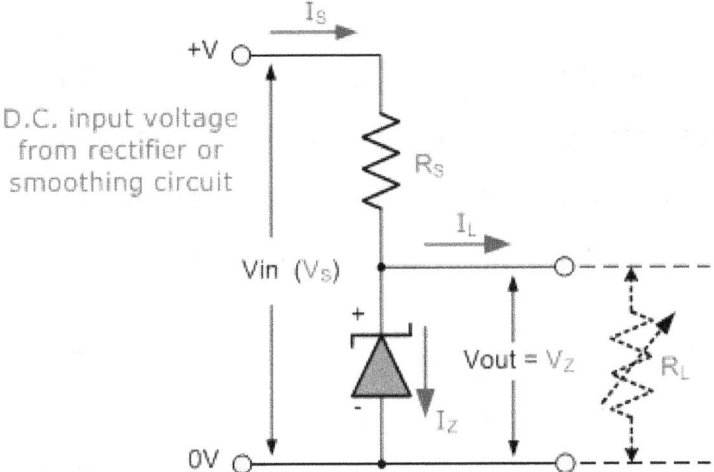

Resistor, R_S is connected in series with the zener diode to limit the current flow through the diode with the voltage source, V_S being connected across the combination. The stabilized output voltage V_{out} is taken from across the zener diode.

The zener diode is connected with its cathode terminal connected to the positive rail of the DC supply so it is reverse biased and will be operating in its breakdown condition. Resistor R_S is selected so to limit the maximum current flowing in the circuit.

With no load connected to the circuit, the load current will be zero, ($I_L = 0$), and all the circuit current passes through the zener diode which in turn dissipates its maximum power.

Also a lesser value of the series resistor R_S will result in a greater diode current when the load resistance R_L is connected and large as this will enhance the power dissipation requirement of the diode so care must be taken when selecting the appropriate value of series resistance so that the zener's maximum power rating is not exceeded under this no-load or high-impedance condition.

The load is connected in parallel with the zener diode, so the voltage across R_L is always the same as the zener voltage, ($V_R = V_Z$).

There is a minimum zener current for which the stabilization of the voltage is effective and the zener current must stay above this value operating under load within its breakdown region at all times. The higher limit of current relying on the power rating of the device. The supply voltage V_S must be larger than V_Z.

In zener diode stabilizer circuits, diode can sometimes produce electrical noise on top of the DC supply as it tries to stabilize the voltage. Usually this is not a problem for most applications but the addition of a large value decoupling capacitor across the zener's output may be required to give extra smoothing.

Zener diodes are always reverse biased. Therefore, Zener diodes can be used to design simple voltage regulator circuits to maintain a constant DC output voltage across the load regardless of input voltage fluctuations or load current fluctuations.

The Zener voltage regulator consists of a current limiting resistor RS connected in series with the input voltage VS and a Zener diode connected in parallel with this reverse biased load RL. The regulated output voltage is always selected to be equal to the diode breakdown voltage V_Z.

Example of Zener Diode

A 5.0V stabilized power supply is required to be produced from a 12V DC power supply input source. The maximum power rating P_Z of the zener diode is 2W. Using the zener regulator circuit above calculate:

a). The maximum current flowing through the zener diode.

$$\text{Maximum Current} = \frac{\text{Watts}}{\text{Voltage}} = \frac{2w}{5v} = 400mA$$

b). The minimum value of the series resistor, R_S

$$R_S = \frac{V_S - V_Z}{I_Z} = \frac{12 - 5}{400mA} = 17.5\Omega$$

c). The load current I_L if a load resistor of 1kΩ is connected across the zener diode.

$$I_L = \frac{V_Z}{R_L} = \frac{5v}{1000\Omega} = 5mA$$

d). The zener current I_Z at full load.

$$I_Z = I_S - I_L = 400mA - 5mA = 395mA$$

Zener Diode Voltages

Zener diodes can not only produce a single stabilized voltage output, but can also be placed in series with a regular silicon signal diode to produce a variety of different reference voltage output values, as shown below.

Zener Diodes Connected in Series

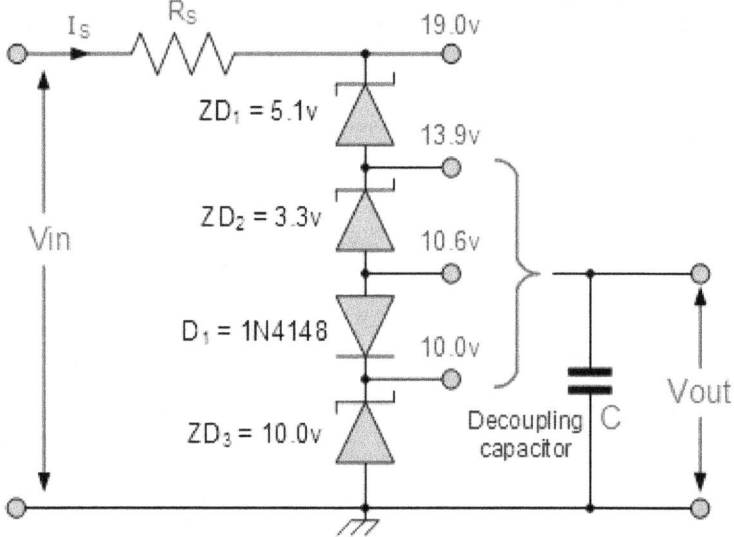

The value of the individual Zener diode can be selected according to the application, but the silicon diode always drops by about 0.6 to 0.7 V in the forward voltage state. Of course, the supply voltage Vin must be greater than the maximum output reference voltage, which is 19V in the above example.

Typical Zener diodes for common electronic circuits are the 500mWBZX55 series or the larger 1.3WBZX85 series, for example, the Zener voltage of a 7.5V diode is designated as C7V5. This is the diode reference number for the BZX55C7V5.

500mW series Zener diodes are available at about 2.4 to about 100 volts and usually have the same sequence of values as the 5% (E24) resistor series, with individual voltage ratings for these small but very useful diodes.

Zener Diode Standard Zener Voltages

BZX55 Zener Diode Power Rating 500mW							
2.4V	2.7V	3.0V	3.3V	3.6V	3.9V	4.3V	4.7V

5.1V	5.6V	6.2V	6.8V	7.5V	8.2V	9.1V	10V
11V	12V	13V	15V	16V	18V	20V	22V
24V	27V	30V	33V	36V	39V	43V	47V

BZX85 Zener Diode Power Rating 1.3W

3.3V	3.6V	3.9V	4.3V	4.7V	5.1V	5.6	6.2V
6.8V	7.5V	8.2V	9.1V	10V	11V	12V	13V
15V	16V	18V	20V	22V	24V	27V	30V
33V	36V	39V	43V	47V	51V	56V	62V

Zener Diode Clipping Circuits

Diode clipping and clamping circuits are circuits used to shape or modify the AC input waveform (or any sine wave) and produce differently shaped output waveforms depending on the circuit layout. The diode clipper circuit is also called a limiter because it limits or clips the positive (or negative) part of the AC input signal. Zener clipper circuits are primarily used in circuit protection or waveform shaping circuits because they limit or clip parts of the waveform. For example, if you want to clip the output waveform at + 7.5V, use a 7.5V Zener diode. When the output waveform tries to exceed the 7.5V limit, the Zener diode "clip" the overvoltage from the input, creating a flat top waveform that keeps the output constant at + 7.5V. Zener diodes are still forward biased diodes, and when the AC waveform output goes negative below 0.7V, the Zener diode turns "on" like a normal silicon diode, with an output of 0.7V as shown below:

Square Wave Signal

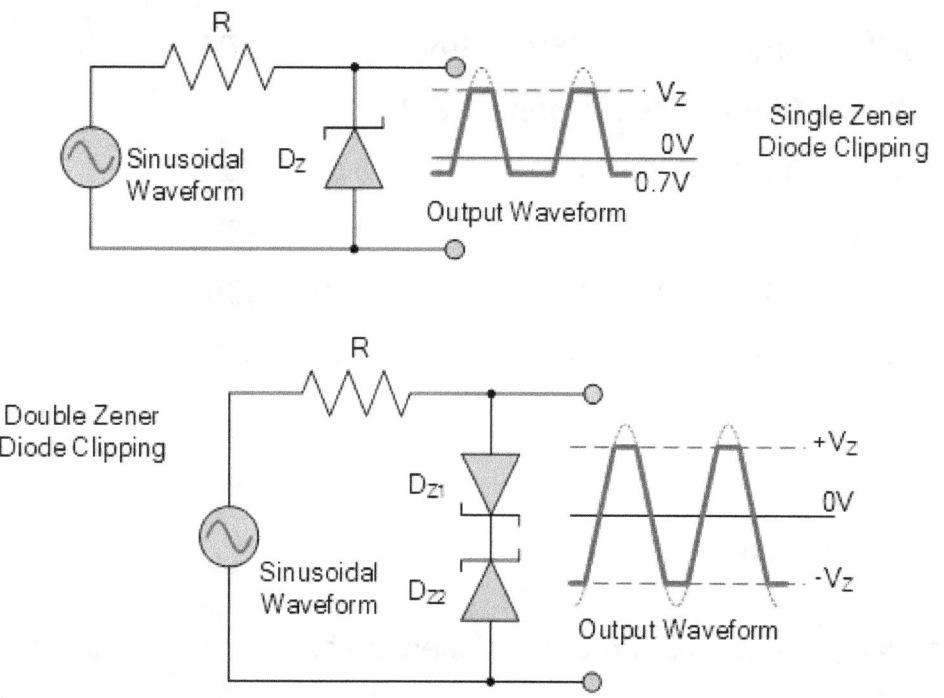

The Zener diode which has been connected back to back can be used as an AC regulator, creating what is known as the "poor man's square wave generator". This arrangement allows you to clip the waveform between the positive + 8.2V and negative 8.2V values of the 7.5V Zener diode. For example, if you want to clip the output waveform between two different minimum and maximum values (for example, + 8V and 6V), you just need to use two differently rated Zener diodes. Note that when the diode voltage is applied in the forward direction, the output will actually clip the AC waveform from + 8.7V to 6.7V. In other words, the forward bias voltage drop across the diode adds another 0.7 volts in each direction, resulting in a peak-to-peak voltage of 15.4 volts instead of the expected 14 volts. This type of clipper configuration is very common for protecting electronic circuits from overvoltage. The two Zeners are usually located between the input terminals of the power supply, and during normal operation, one of the Zener diodes is "off" and the diode has little or no effect. However, when the input voltage waveform exceeds the limit, the Zener turns "on" and clips the input to protect the circuit. In the next tutorial on diodes, we'll look at how to use a diode's forward-biased PN junction to produce light. Previous tutorials have shown that as carriers move across a junction, electrons combine with holes and energy is lost in the form of heat, although some of that energy is dissipated as photons. They are invisible. Placing a translucent lens around the junction produces visible light and the diode is the light source. This effect creates another type of diode, commonly referred to as a light emitting diode. This diode uses this light-generating property to emit light (photons) of different colors and wavelengths.

CHAPTER-8: THE LIGHT EMITTING DIODE

The Light Emitting Diode

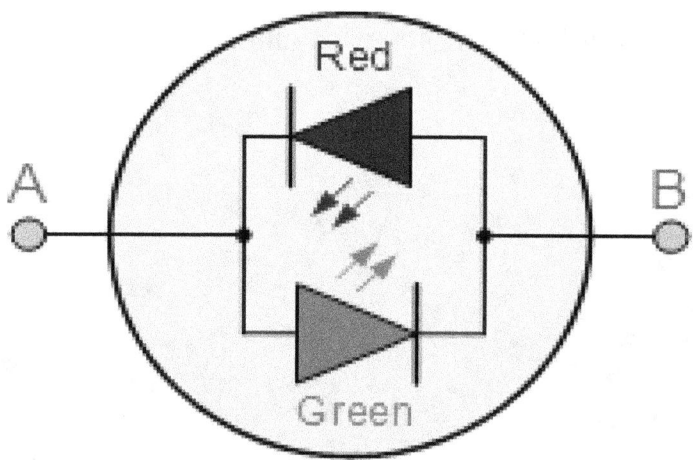

LEDs (Light Emitting Diodes) are semiconductor devices that convert electrical energy into light energy. LEDs are one of the most common types of semiconductor diodes available today and are commonly used in televisions and color displays. They are the most visible type of diode, emitting a fairly narrow range of visible light of different color wavelengths, invisible infrared light for remote controls, or laser light when a forward current flows. More commonly referred to as "light emitting diodes" or LEDs are basically special types of diodes because they have electrical characteristics very similar to PN junction diodes. This means that the LED will carry current in the forward direction and block the current in the reverse direction. The light emitting diode is composed of a very thin layer of semiconductor material doped to a fairly high concentration, and when biased forward depending on the semiconductor material used and the amount of doping, the LED will have a specific spectral wavelength. It emits colored light. When the diode is forward biased, the electrons from the conduction band of the semiconductor recombine with the holes from the valence band, emitting enough energy to generate photons that emit monochromatic (monochromatic) light. increase. Due to this thin layer, a reasonable number of these photons can radiate out of the junction and create a colored light output.

The Structure of LED

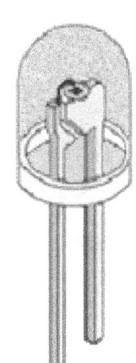

The structure of a light emitting diode is very different from that of a normal signal diode. The LED PN junction is surrounded by a transparent, hard epoxy hemispherical shell or body that protects the LED from both vibration and shock. Surprisingly, LED junctions don't really emit that much light. Therefore, the photons of the light emitted from the junction are reflected off the surrounding board base where the diode is mounted, and the dome-shaped top of the LED itself acts like a lens, concentrating the amount of light. increase. For this reason, the emitted light appears brightest at the top of the LED. However, not all LEDs are made of hemispherical domes for epoxy envelopes. Some indicator LEDs have a rectangular or cylindrical structure with a flat top surface, or the body is in the shape of a bar or arrow. In general, all LEDs are made up of two legs protruding from the bottom of the body. Also, in almost all modern light emitting diodes, the cathode terminal (-) is characterized by a longer body notch or flat spot, or anode lead (+), resulting in a shorter cathode lead than the others. From the cathode (k). Unlike ordinary incandescent and incandescent bulbs, which generate a lot of heat when illuminated, light emitting diodes produce a "cold light" output, and most of the energy produced is emitted in the visible spectrum, so it is "incandescent". It is more efficient than a light bulb. Because LEDs are solid-state devices, they are extremely small and durable, providing a much longer lamp life than regular light sources.

Light Emitting Diode Colors

Unlike conventional signal diodes designed for power sensing or rectification and made from germanium or silicon semiconductor materials, light emitting diodes are made from exotic semiconductor compounds such as gallium arsenide (GaAs), gallium phosphide (GaP), gallium arsenide phosphide (GaAsP), silicon carbide (SiC) or gallium-indium nitride (GaInN) are all mixed in different proportions to produce distinct color wavelengths.

Different LED compounds emit light in specific regions of the visible light spectrum and thus produce different intensity levels. The exact choice of the semiconductor material used determines the overall wavelength of the photonic light emission and thus produces

Typical LED Characteristics			
Semiconductor Material	Wavelength	Colour	V_F @ 20mA
GaAs	850-940nm	Infra-Red	1.2v
GaAsP	630-660nm	Red	1.8v
GaAsP	605-620nm	Amber	2.0v
GaAsP:N	585-595nm	Yellow	2.2v
AlGaP	550-570nm	Green	3.5v
SiC	430-505nm	Blue	3.6v
GaInN	450nm	White	4.0v

Therefore, the actual color of the light emitting diode is determined by the wavelength of the emitted light and then by the actual semiconductor compound used to form the PN junction during manufacturing. Therefore, the color of the light emitted by the LED is not determined by the color of the LED's plastic body, but these are slightly colored to improve the light output and show the color when not illuminated by a power source. Light emitting diodes come in a variety of colors, the most common being RED, AMBER, YELLOW, and GREEN, making them widely used as visual indicators and moving light displays. Recently developed blue and white LEDs are also available, but with the manufacturing cost of mixing two or more complementary colors in the correct ratio within the semiconductor compound, and the nitrogen atom to the crystal structure. From the above table, the main P-type dopant used in the manufacture of light emitting diodes is gallium (Ga, atomic number 31), and the main N-type dopant used is arsenic (As, atomic number 33). I understand. It provides a composite crystal structure of the resulting gallium arsenide (GaAs). The problem with using gallium arsenide alone as a semiconductor compound is that when a forward current flows, it emits a large amount of low-intensity infrared rays (about 850 nm to 940 nm) from the junction. The amount of infrared light it produces is fine for TV remotes, but not very useful when using LEDs as indicator lights. However, by adding phosphorus (P, atomic number 15) as the third dopant, the total wavelength of the emitted radiation is reduced to less than 680 nm, making red light visible to the human eye. Further improvements to the PN junction doping process have provided a visible light spectrum, as well as a range of colors covering infrared and UV wavelengths, as seen above. By mixing different semiconductor, metal and gas compounds, you can create the following list of LEDs:

Types of Light Emitting Diode

- Gallium Arsenide (GaAs) – infra-red
- Gallium Arsenide Phosphide (GaAsP) – red to infra-red, orange

- Aluminium Gallium Arsenide Phosphide (AlGaAsP) – high-brightness red, orange-red, orange, and yellow
- Gallium Phosphide (GaP) – red, yellow and green
- Aluminium Gallium Phosphide (AlGaP) – green
- Gallium Nitride (GaN) – green, emerald green
- Gallium Indium Nitride (GaInN) – near ultraviolet, bluish-green and blue
- Silicon Carbide (SiC) – blue as a substrate
- Zinc Selenide (ZnSe) – blue
- Aluminium Gallium Nitride (AlGaN) – ultraviolet

Similar to traditional PN junction diodes, light emitting diodes are current dependent devices, and their forward voltage drop VF depends on the semiconductor compound (its bright color) and the forward biased LED current. The most common LEDs require a forward operating voltage of approximately 1.2 to 3.6 volts and a forward current rating of approximately 10 to 30 mA, with 12 to 20 mA being the most common range. Both forward voltage and current depend on the semiconductor material used, but the point at which conduction begins and light is generated is about 1.2V for a standard red LED and about 3.6V for a blue LED. Of course, the exact voltage drop will vary from manufacturer to manufacturer due to the different wavelengths of the doping materials used. The voltage drops across the LED at a given current value, e.g. 20mA, also depends on the initial line VF point. Since LEDs are effectively diodes, the forward current-voltage characteristic curve for each diode color can be plotted as follows:

I-V Characteristics of Light Emitting Diodes

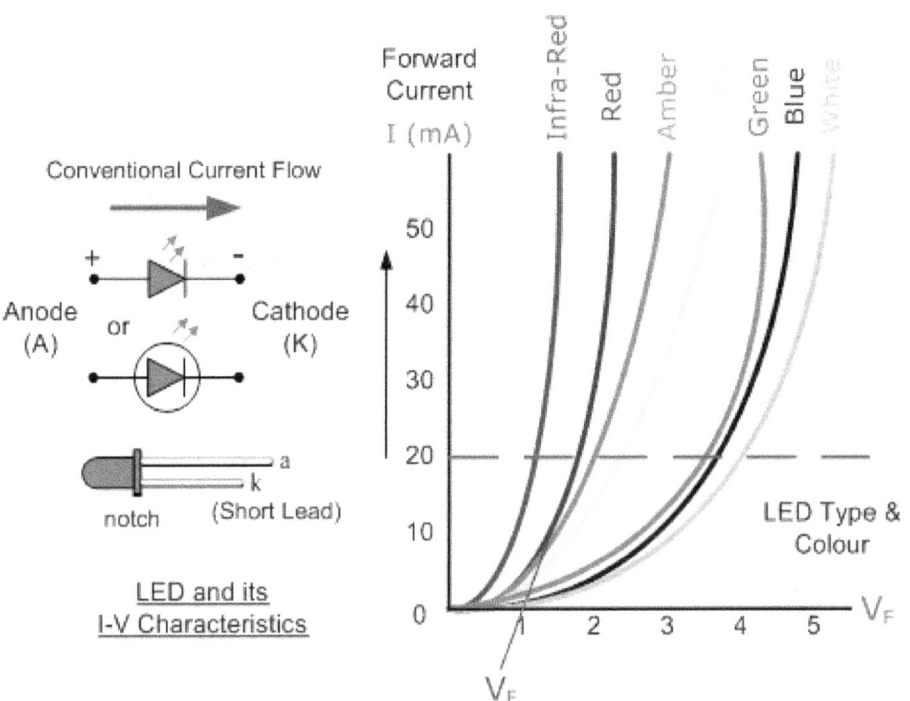

Light emitting diode (LED) schematic symbol and IV characteristics demonstrates the different colors available. Light emitting diodes are current-dependent devices that have a light output intensity that is directly proportional to the forward current flowing through the LED, so they must carry current before they can "emit" any form of light. Since the LEDs are intended to be connected in the forward direction of the power supply, the current must be limited by a series resistance to protect against overcurrent. Do not connect the LED directly to the battery or power supply. The LED is destroyed almost instantly because too much current flows through it and it burns out. From the table above, each LED has its own forward voltage drop across the PN junction, and this parameter, determined by the semiconductor material used, is the forward voltage drop for a given amount of forward line current. You can see that there is. 20mA forward current. In most cases, LEDs are powered by a low voltage DC power supply and use a series resistance RS to limit the forward current to a safety level of, for example, 5mA when high power light output brightness is required.

LED Series Resistance

By using Ohm's Law, we can calculate the series resistor value R_S, by knowing the required forward current I_F of the LED, the supply voltage V_S across the combination and the expected forward voltage drop of the LED, V_F at the required current level, the current limiting resistor is calculated as follows:

LED Series Resistor Circuit

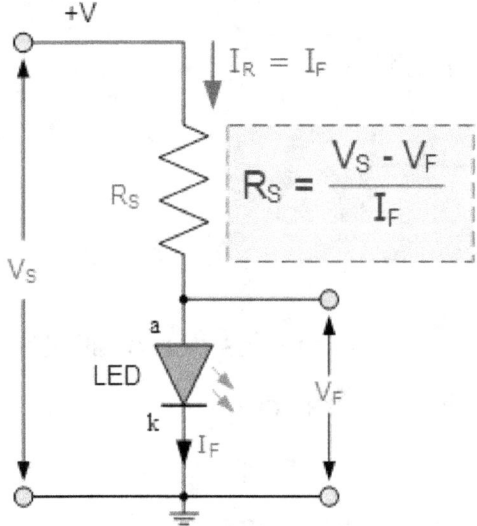

Example of Light Emitting Diode

An LED with yellowish-brown colored having a forward voltage drop of 2 volts connected to a 5.0v steadied DC power supply. Using the circuit above calculate the value of the series resistor required to limit the forward current to less than

10mA. Also calculate the current flowing through the diode if a 100Ω series resistor is used instead of the calculated first.

1). series resistor required at 10mA.

$$R_S = \frac{V_S - V_F}{I_F} = \frac{5v - 2v}{10mA} = \frac{3}{10 \times 10^{-3}} = 300\Omega$$

2). with a 100Ω series resistor.

$$R_S = \frac{V_S - V_F}{I_F}$$

$$\therefore I_F = \frac{V_S - V_F}{R_S} = \frac{5-2}{100} = 30mA$$

We know that resistors come in standard preferred values. Our first calculation above shows that in order to limit the current flowing through the LED to 10mA exactly, a 300Ω resistor is required. In the E12 series of resistors there is no 300Ω resistor. Therefore, we should choose the next uppermost value, which is 330Ω. A quick re-calculation demonstrate that the new forward current value is now 9.1mA, and this is ok.

Connecting LEDs Together in Series

In order to increase required number or to increase the light level when used in displays, LED's can be connected together in series. As with series resistors, LED's connected in series all have the same forward current, I_F flowing through them as just one. As all the LEDs connected in series pass the same current it is usually best if they are all of the same category or color.

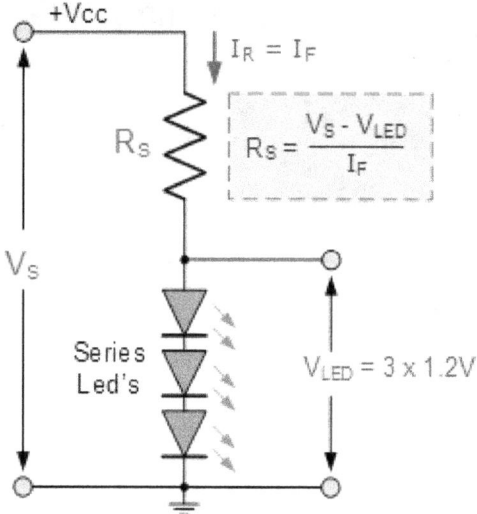

Even though the LED series chain has the same current flowing through it, the series voltage drop across them requires to be considered when calculating the required resistance of the current limiting resistor, R_S. Let's consider that each LED has a voltage drop across it when illuminated of 1.2 volts. In this case, the voltage drop across all three will be 3 x 1.2v = 3.6 volts.

Furthermore, consider that the three LEDs are to be illuminated from the same 5 volt logic device or supply with a forward current of about 10mA, the same as above. Then the voltage drop across the resistor, R_S and its resistance value will be calculated as follows:

$$V_{LED} = 3 \times 1.2 \text{volts} = 3 \times 1.2v = 3.6v$$

$$R_S = V_S - V_{LED} = 5 - 3.6 = 1.4 \text{volts}$$

$$\therefore R_S = \frac{1.4v}{10mA} = 140\Omega$$

Again, in the E12 (10% tolerance) series of resistors there is no 140Ω resistor so we would need to choose the next highest value, which is 150Ω.

LED Driver Circuits

Now that we know what an LED is, we need a way to control it by switching it "on" and "off". Both TTL and CMOS logic gate output stages can be used to drive LEDs because they can source and sink useful amounts of current. A typical integrated circuit (IC) has a maximum output drive current of 50mA in a sink mode configuration, but an internal limiting output current of approximately 30mA in a source mode configuration.

We observed that the LED current should always be limited to a safe value with series resistance. Below are some examples of drive light emitting diodes with inverting ICs, but the same idea applies to any type of integrated circuit output, either combined or consecutive.

IC Driver Circuit

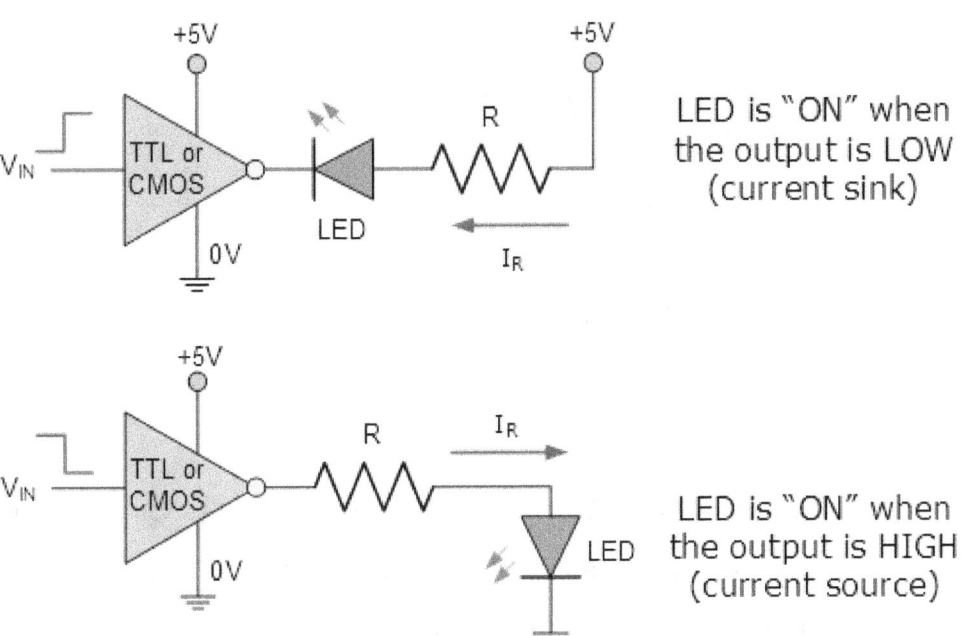

If we need to drive multiple LEDs at the same time like large LED arrays, or if the load current is too high for an integrated circuit, or if we want to use separate components instead of ICs, then a substitute technique to drive the LEDs using either bipolar NPN or PNP transistors as switches is given below. Again as before, a series resistor, RS is necessary to limit the LED current.

Transistor Driver Circuit

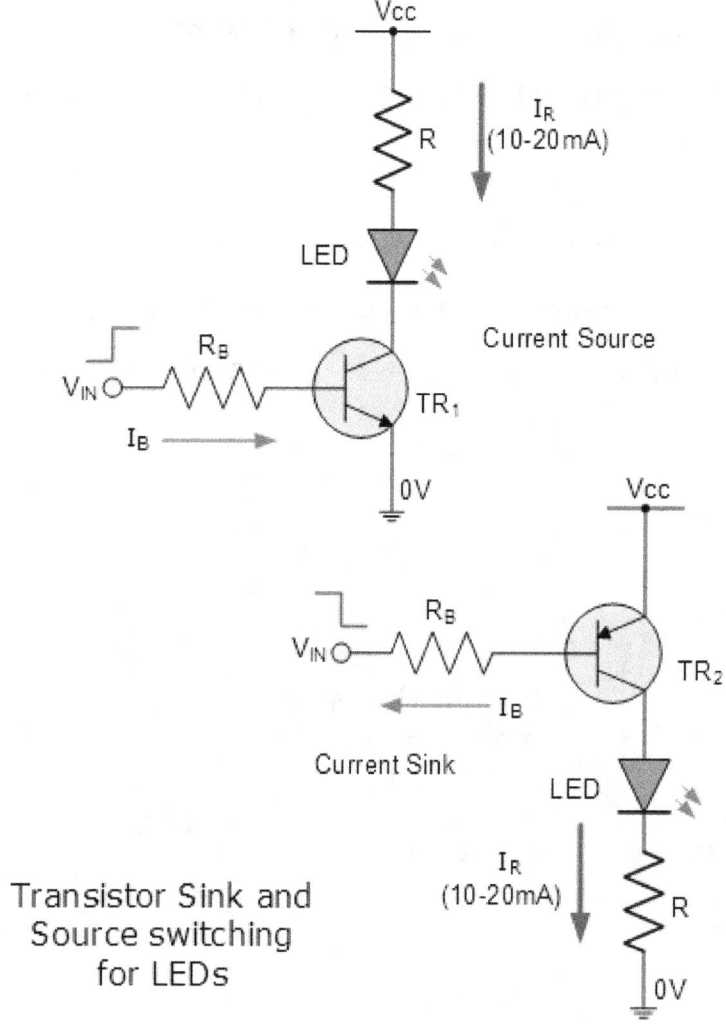

Transistor Sink and Source switching for LEDs

The brightness of a light-emitting diode cannot be regulated by varying the current flowing through it. Allowing more current to flow through the LED will make it brighter, but will also cause it to dissipate more heat. LEDs are designed to produce an active amount of light at a specific forward current between 10 and 20 mA. In situations where saving energy is important, less current may be available. However, dropping the current to less than 5mA, for example, could dim its light output too much, or even turn off the LED entirely. A much better way to control LED brightness is to use a control process known as "pulse width modulation" or PWM, where the LED is switched on and "off" in cycles over and over. at different frequencies depending on the required light intensity of the LED.

LED Light Intensity using PWM

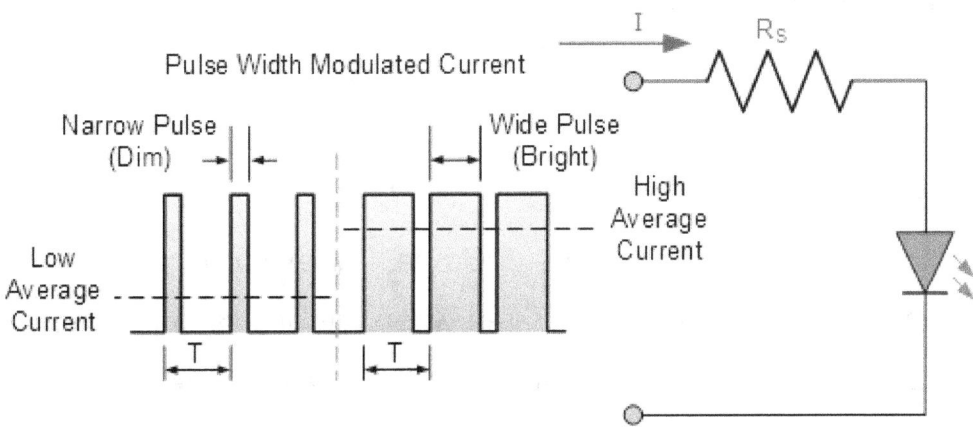

If higher light output is necessary, a pulse width modulation current with a relatively short duty cycle ("ON-OFF" ratio) will significantly increase the diode current, and thus the output light intensity, during the actual pulse while the LED "average". Current level "and power loss within safety limits.

This "ON-OFF" blinking state is what the human eye sees because it "fills" the gap between the "ON" and "OFF" optical pulses when the pulse frequency is high enough to appear stable. Does not affect. The light is displayed as the light output. Therefore, pulses with frequencies above 100 Hz actually appear brighter to the eye than continuous light of the same average intensity.

Multi-colored Light Emitting Diode

LEDs come in different shapes, colors, and sizes, and have different light output intensities. The most common (and cheapest to manufacture) is the standard 5mm red gallium arsenide phosphorus (GaAsP) LED. LEDs are also available in various "packages" arranged to generate both letters and numbers. The "7-segment display" arrangement is the most common. Currently, full-color LED flat panel displays, mobile phones and televisions are available, each using a large number of multi-color LEDs driven directly by a dedicated IC. Most light emitting diodes produce only a monochromatic light output, but multicolor LEDs are now available that can produce different colors on a single device. Most of these are actually two or three LEDs made in one package.

Two-color Light Emitting Diodes

A two-color light emitting diode is a package of two "antiparallel" (one forward and one reverse) LED chips connected in one package. For example, a two-color LED can produce one of three colors. It emits red when the device is connected to a current in one direction and green when it is biased in the other direction.

This type of bidirectional placement helps to show polarity. Correct connection of batteries, power supplies, etc. Also, if the device is connected to a low voltage, low frequency AC power source (via the appropriate resistance), the two LEDs will light alternately, so the bidirectional current will mix both colors.

A Bi-color LED

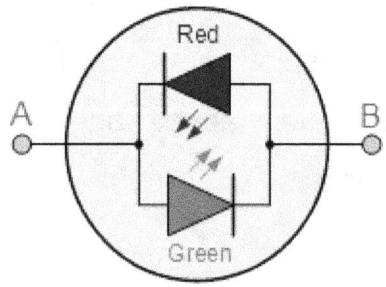

LED Selected	Terminal A		AC
	+	–	
LED 1	ON	OFF	ON
LED 2	OFF	ON	ON
Color	Green	Red	Yellow

Tricolored Light Emitting Diode

The most popular type of 3-color light emitting diode combines a single red and green LED into one package and combines the cathode terminals to create a 3-terminal device. These are called three-color LEDs because you can emit a single red or green color by turning on only one LED at a time. These three-color LEDs are primary colors such as orange and yellow (third color) by turning the two LEDs "on" with different forward current ratios to produce four different colors, as shown in the table. You can also generate additional shades of. From two diode junctions. The most popular type of 3-color light emitting diode combines a single red and green LED into one package and combines the cathode terminals to create a 3-terminal device. These are called three-color LEDs because you can emit a single red or green color by turning on only one LED at a time. These three-color LEDs are primary colors such as orange and yellow (third color) by turning the two LEDs "on" with different forward current ratios to produce four different colors, as shown in the table. In this way, we can generate additional shades by using two diode junctions.

A Multi or Tricolored LED

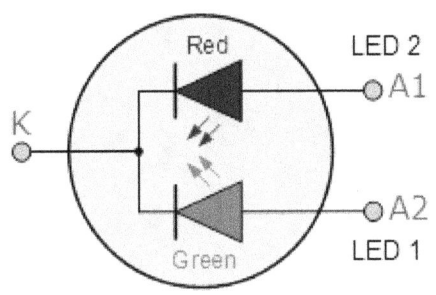

Output Colour	Red	Orange	Yellow	Green
LED 1 Current	0	5mA	9.5mA	15mA
LED 2 Current	10mA	6.5mA	3.5mA	0

LED Displays

In addition to monochromatic or multicolor LEDs, you can combine multiple light emitting diodes into a single package to create displays such as bar graphs, strips, arrays, and 7-segment displays. The 7-segment LED display provides a very convenient way to display information and digital data in the form of numbers, letters and even alphanumerical when properly decoded. As the name implies, they consist of seven separate LEDs (segments). A single ad package. The correct combination of LED segments must be lit on the display to generate the required number or letter from 0 to 9 or A to F. A standard 7-segment LED display typically has eight input ports. One is for each LED segment and the other acts as a common port or port for all internal segments.

> Common Cathode Display (CCD): The Common Cathode Display brings together all the cathode connections of an LED and applies a HIGH logic "1" signal to light individual segments.
> Common Anode Display (CAD): In a common anode display, all anode terminals of an LED are interconnected and each segment is lit by connecting the terminals to a LOW logical "0" signal.

A Typical Seven Segment LED Display

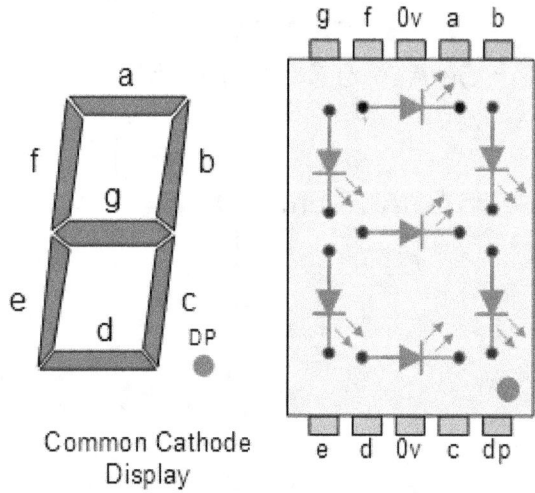

Common Cathode Display

Opto-coupler

An opto-coupler or opto-isolator as it is also called, is a single electronic device that consists of a light emitting diode combined with either a photo-diode, photo-transistor or photo-triac to provide an optical signal path between an input connection and an output connection while maintaining electrical isolation between two circuits.

An opto-isolator consists of a light proof plastic body that has a typical breakdown voltage between the input (photo-diode) and the output (photo-transistor) circuit of up to 5000 volts. This electrical isolation is especially useful where the signal from a low voltage circuit such as a battery powered circuit, computer or microcontroller, is required to operate or control another external circuit operating at a potentially dangerous mains voltage.

Photo-diode and Photo-transistor Opto-couplers

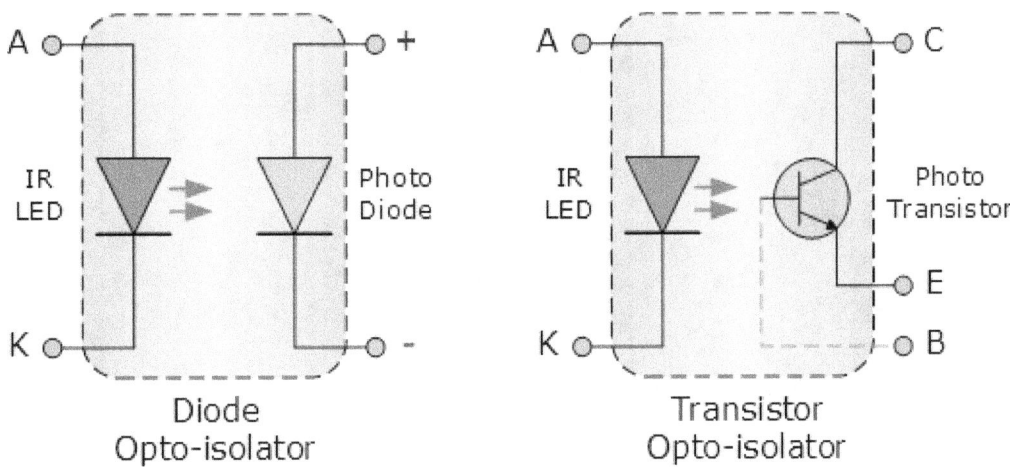

The two components used in an opto-isolator, an optical transmitter like gallium arsenide LED that emit infrared light and an optical receiver as a phototransistor are optically coupled and use light to send signals. signal and/or information between its input and output. This allows information to be transferred between circuits without the need for electrical connections or common ground potential.

Opto-isolators are switching or digital devices, so they transmit "ON-OFF" control signals or digital data. The analog signal can be transmitted by either pulse width or frequency modulation.

CHAPTER-9: BYPASS DIODES IN SOLAR PANELS

Bypass Diodes in Solar Panels

In order to provide a current path around solar cells in the event that a cell or panel becomes faulty or open-circuited, Bypass Diodes are connected in parallel with individual solar cells or panels. Bypass diodes are connected in reverse bias between the positive and negative output terminals of a solar cell (or panel) and have no influence on the output.

In an ideal world, each solar cell would have its own bypass diode, but this can be costly, thus one diode is used per small group of series cells. Individual sun cells are used to make a "solar panel," and solar cells are built of layers of silicon semiconductor materials. An overabundance of electrons is created by treating one layer of silicon with a chemical. The negative or N-type layer is formed as a result of this. Similar to transistors and diodes, the other layer is treated to generate an electron shortage and becomes the positive or P-type layer. This silicon structure forms a light-sensitive PN-junction semiconductor when it is combined with conductors.

In actuality, photovoltaic solar cells, or PVs for short, are nothing more than large, flat photosensitive diodes. Without any moving or mechanical parts, photovoltaic solar cells transform photon light around the PN-junction directly into electricity. PV cells get their energy from sunshine rather than heat. In fact, while they're cold, they're most effective!

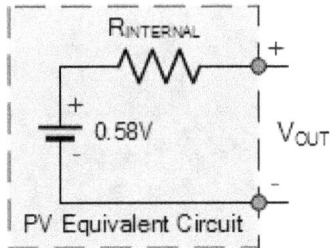

The equivalent circuit of a PV is that of a battery having a series internal resistance, RINTERNAL, similar to any other conventional battery, as shown above. The cell voltage and hence available current will vary amongst photovoltaic cells of equal size and structure, connected to the same load, and under the same light source due to changes in internal resistance, so this must be accounted for in the solar panel assembly you buy.

The photovoltaic solar cell's silicon wafer, which faces the sun, has the electrical connections and is coated with an anti-reflective coating that aids in the effective absorption of sunlight. The semiconductor material and the external electrical load, such as a light bulb or a battery, are connected by electrical connections.

Photons of light strike the surface of the semiconductor material and free electrons from their atomic bonds when sunlight shines on a photovoltaic cell. Certain doping chemicals are added to the semiconductor's composition during manufacturing to aid in the establishment of a channel for the released electrons.

These routes cause an electrical current to flow over the surface of the photovoltaic solar cell, creating a flow of electrons. A photovoltaic cell's surface is covered with metallic strips that gather the electrons that make up the cell's positive (+) connection. The negative (−) connection to the cell is formed by a layer of aluminum or molybdenum metal on the back of the cell, away from the incoming sunlight.

Construction of Photovoltaic Solar Cell

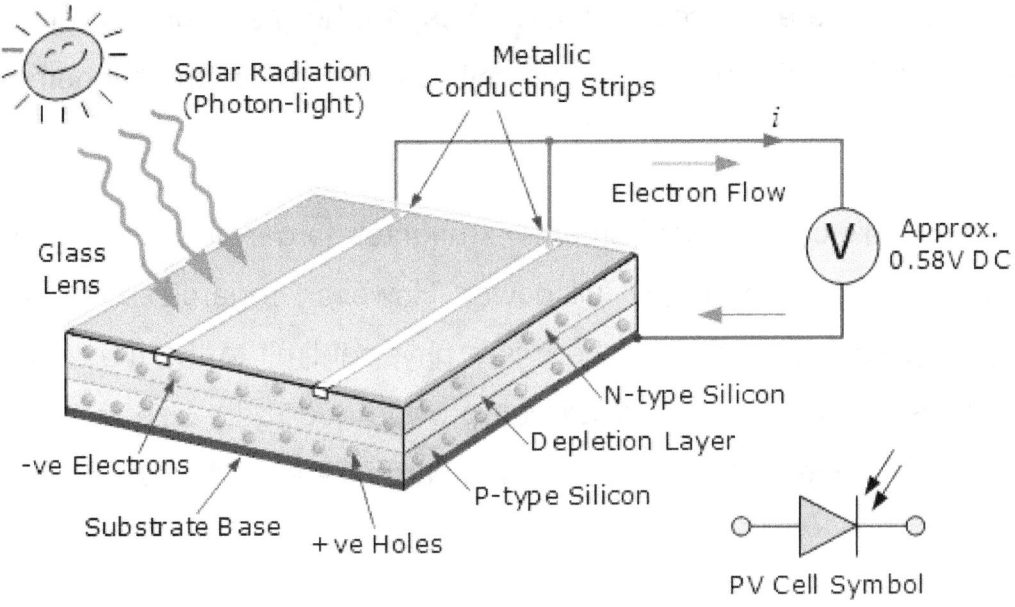

Photovoltaic (PV) solar cells produce DC power when exposed to sunlight, just like a battery or cell. Most photovoltaic solar cells produce a maximum "no-load" open circuit voltage (V_{OUT}) of roughly 0.5 to 0.6 volts when no external circuit or load is attached to its terminals, which is substantially less than a conventional 1.5V dry battery cell. Higher voltages can be generated by connecting a number of PV cells in series, just like batteries.

A photovoltaic cell produces a current (I) proportional to the amount of sunlight falling on its surface when exposed to sunshine. The maximum current a PV cell can produce, known as its short-circuit current ISC, occurs when the terminals of the cell are shorted together, but under these maximum current conditions, the terminal voltage is zero, $V_{OUT} = 0$.

The output voltage of a photovoltaic cell is then highly dependent on the load current demands from ISC to IO. A PV cell is thus essentially a low-voltage, high-current device. The current (and power) output of a photovoltaic cell is proportional to the amount of sunlight striking the cell's surface. For example, cloudy or dull days reduce a PV cell's effectiveness, so the maximum current it can supply to a given load is reduced, but the cell can still provide the full output voltage.

For increasing the load's current requirements, a brighter, larger amount of solar radiation would be required to supply full power. However, regardless of how intense or bright the sun's radiation is, there is a physical limit to the maximum current that a single photovoltaic solar cell can provide due to its size (surface area). This is known as the maximum deliverable current and is denoted by the symbol I_{MAX}.

The I_{MAX} value of a single photovoltaic solar cell is determined by the cell's size or surface area (especially the PN-junction), the amount of direct sunlight hitting the cell, the efficiency of converting this solar power into a current, and, of course, the type of semiconductor material used to manufacture the cell (silicon, gallium arsenide, cadmium Sulphur, or cadmium telluride, for example). When selecting blocking diodes or bypass diodes to connect to solar cells or panels, this maximum current value, I_{MAX}, must be considered.

Diodes in Photovoltaic Arrays

The PN-junction diode functions like a solid-state one-way electrical valve, allowing electrical current to flow in only one direction. The advantage of this is that diodes can be used to prevent electric current from flowing from other parts of an electrical solar circuit. When used in conjunction with a photovoltaic solar panel, these silicon diodes are referred to as Blocking Diodes.

Bypass diodes are used in parallel with a single or multiple photovoltaic solar cells to prevent currents flowing from good, well-exposed to sunlight solar cells from overheating and burning out weaker or partially shaded solar cells by providing a current path around the bad cell. Blocking diodes are not used in the same way as bypass diodes.

In solar panels, bypass diodes are connected "parallel" to a photovoltaic cell or panel to shunt current around it, whereas blocking diodes are connected "series" to the PV panels to prevent current from flowing back into them. Blocking diodes differ from bypass diodes because, while the diode is physically the same in most cases, it is installed differently and serves a different purpose. Consider the solar photovoltaic array shown below.

Bypass Diodes in Photovoltaic Arrays

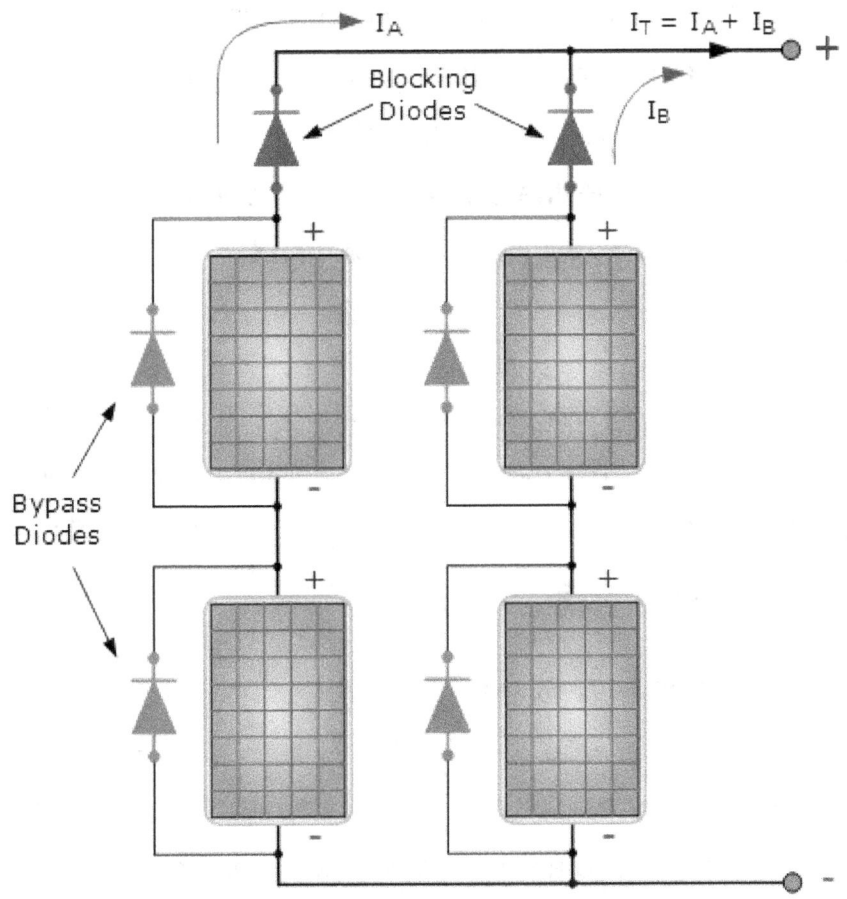

As previously stated, diodes are devices that allow current to flow in only one direction. The green diodes shown above are "bypass diodes," which are connected in parallel with each solar panel to provide a low resistance path. This short circuit current must be safely carried by bypass diodes in solar panels and arrays. The two red diodes, one in series with each series branch, are referred to as "blocking diodes." Although blocking diodes differ from bypass diodes, the two diodes are physically identical in most cases. They are, however, installed differently and serve a different purpose.

These blocking diodes, also known as series diodes or isolation diodes, ensure that electrical current flows in only one direction "OUT" of the series array, to the external load, controller, or batteries. This is done to prevent current generated by other parallel-connected PV panels in the same array from flowing back through a weaker (shaded) network, as well as to prevent fully charged batteries from discharging or draining back through the array at night. When connecting multiple solar panels in parallel, blocking diodes should be used in each parallel connected branch.

Blocking diodes are typically used in PV arrays when there are two or more parallel branches or when some of the array may become partially shaded during the day as the sun moves across the sky. The size and type of blocking diode used are determined by the photovoltaic array.

As bypass diodes in solar panels and arrays, two types of diodes are available: PN-junction silicon diodes and Schottky barrier diodes. Both are available in a variety of current ratings. The Schottky barrier diode has a much lower forward voltage drop of about 0.4 volts compared to a silicon device's 0.7 volt drop.

This lower voltage drop allows the solar array to save one full PV cell in each series branch, making the array more efficient because less power is dissipated in the blocking diode. Most solar panel manufacturers include both blocking and bypass diodes in their panels to simplify the design.

CHAPTER-10: DIODE CLIPPING CIRCUITS

Diode Clipping Circuits

The Diode Clipper, also known as a Diode Limiter, is a wave shaping circuit that clips or cuts off the top, bottom, or both halves of an input waveform. Clipping the input signal results in an output waveform that looks like a flattened version of the input. The half-wave rectifier, for example, is a clipper circuit because all voltages below zero are eliminated.

However, Diode Clipping Circuits can be used in a variety of applications, such as modifying an input waveform with signal and Schottky diodes, or providing over-voltage protection with zener diodes to ensure that the output voltage never exceeds a certain level, protecting the circuit from high voltage spikes. Then, in voltage limiting applications, diode clipping circuits can be used.

In the Signal Diodes tutorial, we learned that when a diode is forward biased, it allows current to flow through it while clamping the voltage. When a diode is reverse biased, no current flows through it and the voltage across its terminals remains unchanged; this is the fundamental operation of the diode clipping circuit. Although the input voltage to diode clipping circuits can have any waveform shape, we will assume it is sinusoidal here. Consider the following circuits.

Positive Diode Clipping Circuits

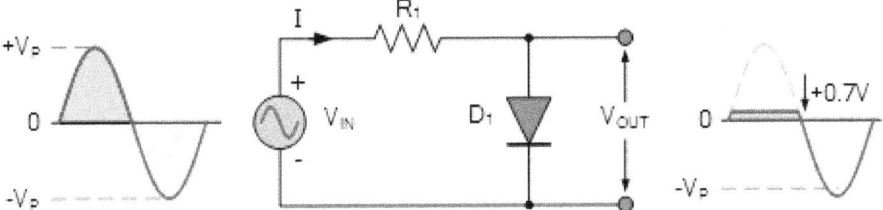

During the positive half cycle of the sinusoidal input waveform, the diode is forward biased (anode more positive than cathode). The input voltage must be greater than +0.7 volts for the diode to become forward biased (0.3 volts for a germanium diode). When this occurs, the diodes begin to conduct and maintain a constant voltage across themselves of 0.7V until the sinusoidal waveform falls below this value. During the positive half cycle, the output voltage taken across the diode can never exceed 0.7 volts.

The diode is reverse biased (cathode more positive than anode) during the negative half cycle, stopping current flow and having no effect on the negative half of the sinusoidal voltage, which goes to the load unaffected. As a result, a positive clipper circuit uses a diode to limit the positive half of the input waveform.

Negative Diode Clipping Circuits

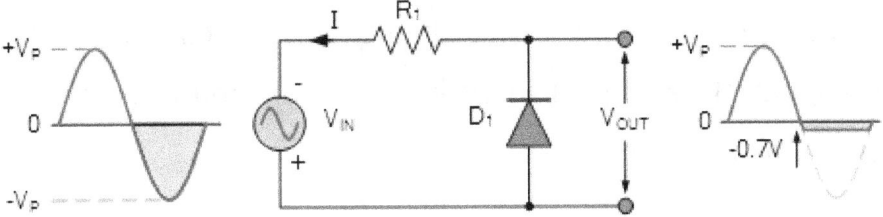

In this case, the reverse is true. When reverse biased, the diode is forward driven during the negative half cycle of the sinusoidal waveform, limiting or clipping it to −0.7 volts while letting the positive half cycle to pass unmodified. Negative clipper circuits are so named because the diode limits the input voltage's negative half cycle.

Clipping of Both Half Cycles

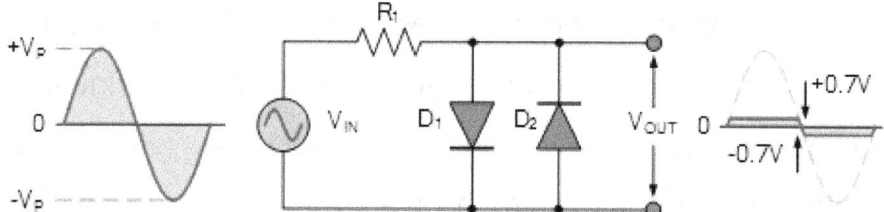

If we connect two diodes in inverse parallel as indicated, both the positive and negative half cycles of the sinusoidal input waveform will be clipped, as diode D1 clips the positive half cycle and diode D2 clips the negative half cycle. Then diode clipping circuits can be used to clip the positive, negative, or both halves of the cycle. The output waveform above would be zero with perfect diodes.

The real clipping point occurs at +0.7 volts and –0.7 volts, respectively, due to the forward bias voltage drop between the diodes. However, we may boost this 0.7V threshold to any value we desire up to the sinusoidal waveform's highest value (VPEAK) by connecting more diodes in series to create multiples of 0.7 volts, or by adding a voltage bias to the diodes.

Biased Diode Clipping Circuits

A bias voltage, VBIAS, is placed in series with the diode to make a combination clipper, as shown, to produce diode clipping circuits for voltage waveforms at different levels. Before the diode gets sufficiently forward biased to conduct, the voltage across the series combination must be larger than VBIAS + 0.7V. If the VBIAS level is set to 4.0 volts, the sinusoidal voltage at the anode terminal of the diode must be larger than 4.0 + 0.7 = 4.7 volts in order for it to become forward biased. Above this bias point, any anode voltage values are cut.

Positive Bias Diode Clipping

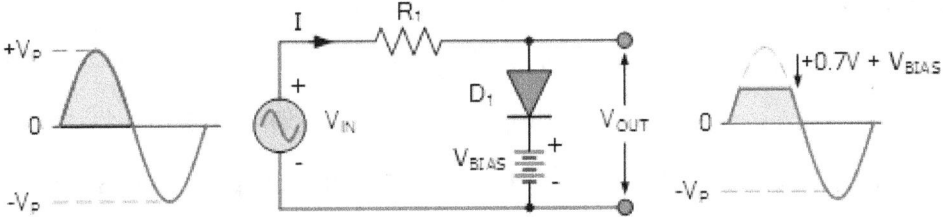

When a diode conducts, the negative half cycle of the output waveform is kept to a level -VBIAS – 0.7V by reversing the diode and battery bias voltage, as shown.

Negative Bias Diode Clipping

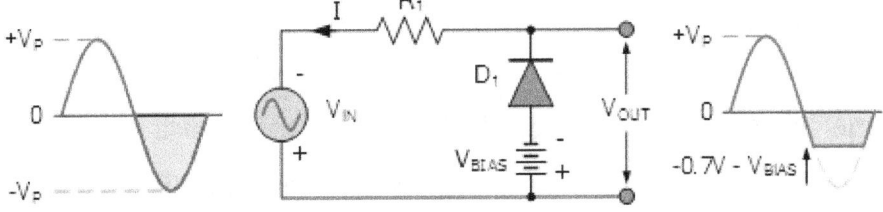

By adjusting the bias voltage of the diodes, a variable diode clipping or diode limiting level can be achieved. Two biased clipping diodes are utilized if both the positive and negative half cycles have to be clipped. The bias voltage does not have to be the same for both positive and negative diode clipping. As depicted, the positive bias voltage could be at one level, such as 4 volts, and the negative bias voltage at another level, such as 6 volts.

Diode Clipping of Different Bias levels

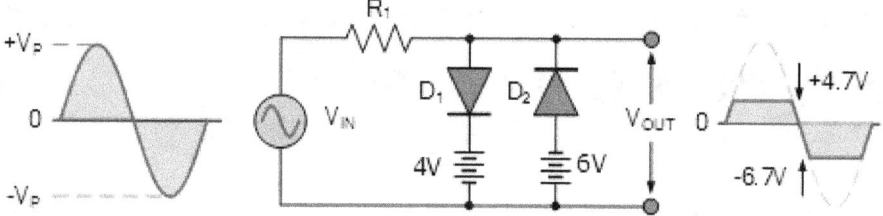

Diode D1 conducts and limits the waveform at +4.7 V when the voltage of the positive half cycle reaches +4.7 V. Until the voltage reaches −6.7 V, diode D2 does not conduct. As a result, any positive voltages greater than +4.7 V and negative voltages less than −6.7 V are clipped automatically. Biased diode clipping circuits provide the advantage of preventing the output signal from exceeding preset voltage limitations for both half cycles of the input waveform, which might be a noisy sensor input or a power supply's positive and negative supply rails. The elimination of both waveform peaks could result in a square-wave shaped waveform if the diode clipping levels are set too low or the input waveform is too large.

Zener Diode Clipping Circuits

The use of a bias voltage allows for precise control of the amount of the voltage waveform that is clipped off. However, one of the major drawbacks of using voltage biased diode clipping circuits is that they require an additional emf battery source, which may or may not be a problem. Zener Diodes are a simple way to create biased diode clipping circuits without the need for an additional emf supply.

The zener diode, as we all know, is a type of diode that has been specially designed to operate in its reverse biased breakdown region and can thus be used for voltage regulation or zener diode clipping applications.

When conducting, the zener behaves like a regular silicon diode in the forward region, with a forward voltage drop of 0.7V (700mV). The voltage is blocked in the reverse bias region until the zener diode's breakdown voltage is achieved. The reverse current through the zener dramatically increases at this time, yet the zener voltage, V_Z across the device remains constant despite the varying zener current, I_Z. Then, as shown, we can put this zener action to good use by clipping a waveform with it.

Zener Diode Clipping

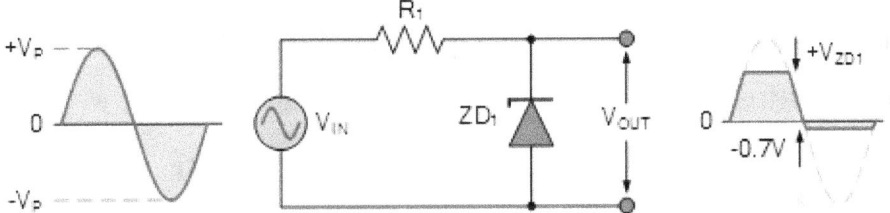

With the bias voltage equal to the zener breakdown voltage, the zener diode acts as a biased diode clipping circuit. The zener diode in this circuit is reverse biased during the positive half of the waveform, therefore the waveform is clipped at the zener voltage, V_{ZD1}. During the negative half cycle, the zener behaves like a regular diode, with a 0.7V junction voltage. Upgrading of this can be done by exploiting the reverse-voltage features of zener diodes to clip both half of a waveform using series connected back-to-back zener diodes, as illustrated.

Full-wave Zener Diode Clipping

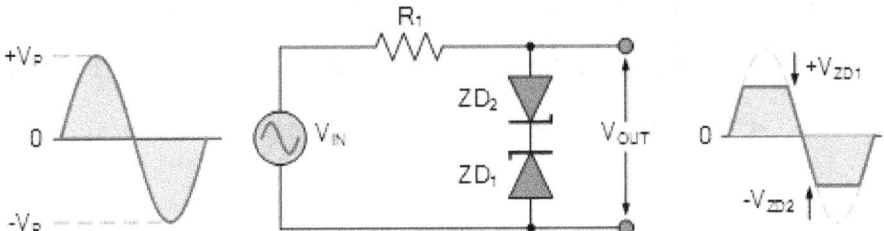

Full wave zener diode clipping circuits produce a waveform that is similar to the voltage biased diode clipping circuit. The output waveform will be truncated at the zener voltage plus the other diode's 0.7V forward volt drop. For example, the positive half cycle will be clipped at the sum of the zener diode, ZD1, plus 0.7V from ZD2, and the negative half cycle will be clipped at the sum of the zener diode, ZD2, plus 0.7V from ZD2.

Zener diodes are available in a variety of voltages and can be used to provide different voltage references on each half cycle, as described above. Zener diodes come in a range of zener breakdown voltages, V_Z, from 2.4 to 33 volts, with a typical tolerance of 1 to 5%. Because full current will flow through the zener diode once it is conducting in the reverse breakdown zone, a suitable current limiting resistor, R1, must be used.

Diode Clipping at a glance

- Diodes can be used to clip the top, bottom, or both of a waveform at a specific dc level and transfer it to the output without distortion, in addition to being rectifiers.
- We've assumed that the waveform is sinusoidal in the examples above, although any curved input waveform can theoretically be utilized.
- Diode Clipping Circuits are used to remove amplitude noise or voltage spikes, regulate voltage, or create new waveforms from an existing signal, such as rounding off the peaks of a sinusoidal waveform to make a rectangular waveform, as seen above.
- The most typical use of "diode clipping" is to protect a switching transistor from reverse voltage transients by connecting a flywheel or free-wheeling diode in parallel across an inductive load.

CHAPTER-11: THE SCHOTTKY DIODE

The Schottky Diode

The Schottky Diode is a metal-semiconductor diode with a very low forward voltage drop and a quick switching speed. The Schottky Diode is another form of semiconductor diode that, like any other junction diode, can be utilized in a variety of wave shaping, switching, and rectification applications.

The key advantage is that a Schottky Diode's forward voltage drop is much lower than a regular silicon pn-junction diode's 0.7 volts. Due to their low power and high switching speeds, Schottky diodes are used in a wide range of applications, including rectification, signal conditioning, and switching, as well as TTL and CMOS logic gates. The letters LS appear somewhere in the logic gate circuit code of TTL Schottky logic gates, such as 74LS00.

When Forward Biased, the depletion region is considerably reduced, allowing current to flow in the forward direction, and when Reverse Biased, the depletion region is expanded, obstructing current flow. The resistance of the junction barrier is decreased or increased by biasing the PN-junction with an external voltage to either forward or reverse bias it. As a result, the resistance value of the junction influences the voltage-current relationship (characteristic curve) of a conventional PN-junction diode.

Because the PN-junction diode is a nonlinear device, its DC resistance varies with the biasing voltage and current flowing through it. Conduction via the junction does not begin until the external biasing voltage reaches the "knee voltage," at which point current rapidly increases, and the voltage necessary for forward conduction to occur in silicon diodes is around 0.65 to 0.7 volts, as indicated.

IV-Characteristics of PN-junction Diode

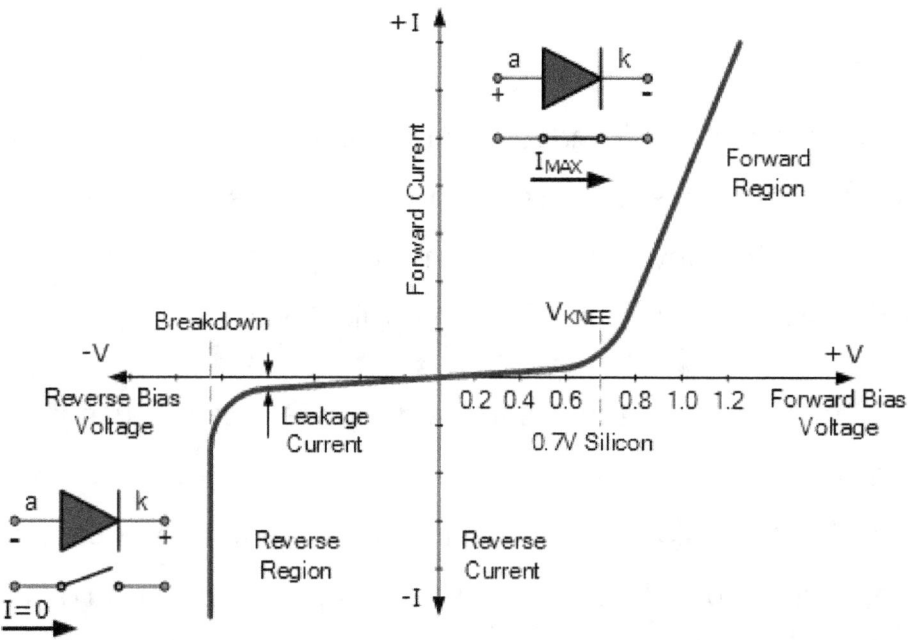

This knee voltage for practical silicon junction diodes can range from 0.6 to 0.9 volts, depending on how the device was doped during manufacturing and whether it's a small signal diode or a much bigger rectifying diode. A normal germanium diode's knee voltage, on the other hand, is substantially lower, at around 0.3 volts, making it better suited to small signal applications.

A Schottky Barrier Diode, or simply "Schottky Diode," is another form of rectifying diode with a low knee voltage and quick switching speed. Schottky diodes can be used in many of the same applications as traditional pn-junction diodes, and they have a wide range of applications, particularly in digital logic, renewable energy, and solar panels.

Construction and Symbol of Schottky Diode

Unlike a traditional pn-junction diode, which is made up of a P-type and an N-type semiconductor, Schottky Diodes are made up of a metal electrode attached to an N-type semiconductor.

Because they are made with a metal compound on one side of their junction and doped silicon on the other, Schottky diodes have no depletion layer and are classified as unipolar devices, as opposed to bipolar pn-junction diodes. "Silicide," a highly conductive silicon and metal compound, is the most common contact metal utilized in Schottky diode manufacturing.

When conducting, this silicide metal-silicon contact has a low ohmic resistance, allowing more current to flow and resulting in a decreased forward voltage V_f drop less than 0.4V. Forward voltage drops are typically between 0.3 and 0.5 volts, depending on the metal composition.

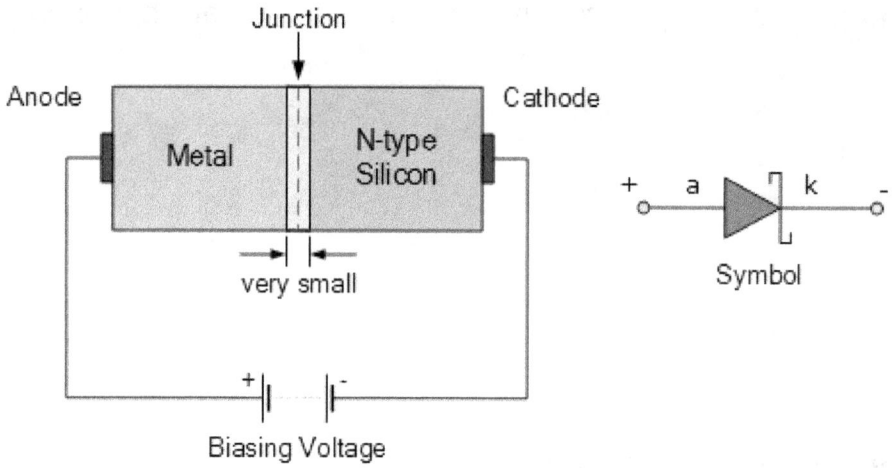

The schematic and symbol for a Schottky diode are shown above, in which a lightly doped n-type silicon semiconductor is linked to a metal electrode to form a "metal-semiconductor junction." The width of this metal-semiconductor junction, and thus its electrical characteristics, will vary greatly depending on the type of metal compound and semiconductor material used in its construction, but when forward-biased, electrons move from the n-type material to the metal electrode, allowing current to flow. The majority carrier drift causes current to flow through the Schottky diode.

When reverse biased, the diode's conduction ceases extremely rapidly and turns to blocking current flow, just like a regular pn-junction diode, because there is no p-type semiconductor material and thus no minority carriers (holes). As a result, a Schottky diode responds to changes in bias relatively quickly, displaying the properties of a rectifying diode.

As previously stated, the knee voltage at which a Schottky diode goes "ON" and begins to conduct is substantially lower than its pn-junction equivalent, as seen in the I-V characteristics below.

IV-Characteristics of Schottky Diode

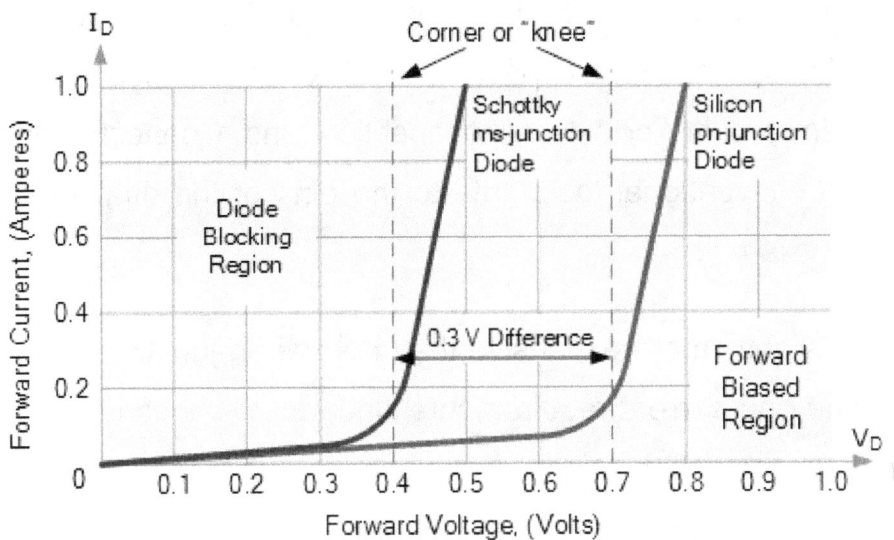

We observed that the I-V characteristics of a metal-semiconductor Schottky diode are remarkably similar to those of a typical pn-junction diode, with the exception that the corner or knee voltage at which the ms-junction diode begins to conduct is significantly lower, at roughly 0.4 volts.

Depending on the metal electrode employed, the forward current of a silicon Schottky diode can be many times greater than that of a standard pn-junction diode due to this lower value.

Remember that power equals volts' times amps (P = V*I), so for a given diode current, I_D, a smaller forward voltage drop will result in less forward power dissipation in the form of heat across the junction. The Schottky diode's decreased power loss makes it an ideal choice for low-voltage, high-current applications like solar photovoltaic panels, where the forward-voltage (VF) drop across a normal pn-junction diode would cause excessive heating.

However, the reverse leakage current (IR) of a Schottky diode is typically substantially higher than that of a pn-junction diode. It should be noted, however, that if the I-V characteristics curve is more linear and non-rectifying, the contact is an Ohmic contact. Ohmic contacts are often utilized to link semiconductor wafers and chips to system circuitry or external connection pins. Connecting the semiconductor wafer of a conventional logic gate to the pins of the dual-in-line (DIL) plastic container, for example.

Schottky diodes are also slightly more expensive than normal pn-junction silicon diodes with similar voltage and current requirements since they are constructed with a metal-to-semiconductor junction. When comparing the 1.0 Ampere 1N58xx Schottky series to the general purpose 1N400x series, for example.

Schottky Diodes in Logic Gates

Due to its improved frequency response, shortened switching times, and lower power consumption, Schottky diodes are widely utilized in Schottky transistor–transistor logic (TTL) digital logic gates and circuits. Schottky-based TTL is the natural solution when high-speed switching is required.

Schottky TTL comes in a variety of forms, each with its own speed and power consumption. The following are the three basic TTL logic series that use the Schottky diode in their construction:

1. **Schottky Diode Clamped TTL (S series):** Schottky "S" series TTL (74SXX) logic gates and circuits are an upgraded version of the original diode-transistor DTL and transistor-transistor 74 series TTL logic gates and circuits. Schottky diodes are used to prevent the base-collector junctions of switching transistors from saturating and causing propagation delays, allowing for faster operation.

2. **Low-Power Schottky (LS series):** The 74LSXX series TTL has improved transistor switching speed, stability, and power dissipation than the prior 74SXX series. The low-power Schottky TTL family consumes less power than the high-speed Schottky TTL family, making the 74LSXX TTL series a good choice for many applications.
3. **Advanced Low-Power Schottky (ALS series):** The 74LSXX series has a shorter propagation delay time and lower power dissipation than the 74ALSXX and 74LS series due to additional advances in the materials used to build the ms-junctions of the diodes. The ALS series is slightly more expensive than ordinary TTL because it is a newer technology with a more complicated internal construction.

Schottky Clamped Transistor

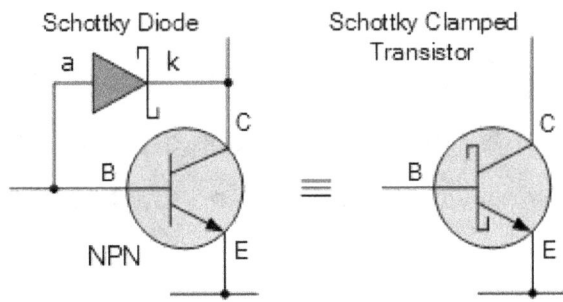

To avoid being forced hard into saturation, all of the prior Schottky TTL gates and circuits use a Schottky clamped transistor. A Schottky clamped transistor is a bipolar junction transistor with a Schottky diode connected in parallel across the base-collector junction, as depicted. The base–collector junction is reverse biased and hence the diode is reverse biased when the transistor conducts properly in the active area of its characteristics curves, allowing the transistor to operate like a conventional npn transistor.

When the transistor begins to saturate, the Schottky diode becomes forward biased and clamps the collector-base junction to its 0.4 volt knee value, preserving the transistor from hard saturation by directing any extra base current through the diode.

Preventing the logic circuits from malfunctioning Schottky TTL circuits are suited for use in flip-flops, oscillators, and memory devices because switching transistors from saturating considerably reduces their propagation delay time.

Schottky Diode at a glance

> - The Schottky Diode, also known as a Schottky Barrier Diode, is a solid-state semiconductor diode in which a metal electrode and an n-type semiconductor create a ms-junction, providing it two important benefits over standard pn-junction diodes: quicker switching speed and low forward bias voltage.
> - For the same amount of forward current, the metal–to–semiconductor or ms-junction offers a substantially lower knee voltage of 0.3 to 0.4 volts, compared to 0.6 to 0.9 volts in a normal silicon base pn-junction diode.
> - Because of differences in the metal and semiconductor materials used in their production, silicon carbide (SiC) Schottky diodes can switch "ON" with a forward voltage drop as low as 0.2 volts, replacing the less common germanium diode in many applications requiring a low knee voltage.

- Schottky diodes are rapidly gaining popularity as the primary rectification technology in low voltage, high current applications such as renewable energy and solar panels.
- Schottky diodes, on the other hand, have higher reverse leakage currents and a lower reverse breakdown voltage (about 50 volts) than pn-junction equivalents.
- The Schottky diode is particularly beneficial in many integrated-circuit applications, with the 74LSXX TTL family of logic gates being the most common. It has a lower turn-on voltage, faster switching time, and lower power consumption.
- By depositing the metal electrode atop severely doped (and consequently low-resistivity) semiconductor areas, metal–semiconductor junctions can be used as "Ohmic contacts" as well as rectifying diodes.
- Ohmic connections conduct current in both directions, making it possible to connect semiconductor wafers and circuits to external terminals.

www.ingramcontent.com/pod-product-compliance
Lightning Source LLC
Chambersburg PA
CBHW060424220526
45465CB00008B/3000